VETERINARY SCIENCE
EXTENSION AND MANAGEMENT

VETERINARY SCIENCE
EXTENSION AND MANAGEMENT

by

Sachin P C

Veena V S

Tina B S

Wordpen Academics

Publishing Science That Matter

ISBN: 978-93-6084-019-8 (Hardback)

Published by : **WORDPEN ACADEMICS**
4736/23, Ansari Road, Darya Ganj,
New Delhi – 110 002
Phone: +91-011-35004336
E-mail: info@wordpenacademics.com

Printed at **:** **Replika Press Pvt. Ltd.**

Exclusive Distributor: ***Bio-Green Books, New Delhi***

Preface

Veterinary Science: Extension and Management provides an in-depth exploration of the critical role veterinary science plays in the health and management of animals, as well as the extension services that support these practices. Designed for students, professionals, and practitioners in the fields of veterinary science, animal husbandry, and agriculture, this book aims to equip readers with comprehensive knowledge on how veterinary practices are extended to rural and underserved populations and how effective management strategies can improve the efficiency of veterinary services.

The book begins by outlining the fundamentals of veterinary science, delving into the principles and practices that ensure animal health, welfare, and disease prevention. It provides a solid foundation in animal physiology, pathology, and clinical practices, focusing on the evolving role of veterinarians in both domestic and wild animal populations. The text also explores the importance of veterinary research and its impact on improving animal health, food safety, and disease control, both at the local and global levels.

A key component of this book is its focus on veterinary extension services. It highlights the importance of communication, education, and outreach programs aimed at transferring veterinary knowledge to farmers, livestock owners, and rural communities. It addresses the challenges and opportunities in providing veterinary care in rural and remote areas, emphasizing the significance of extension workers in bridging the gap between veterinary research and practical applications in the field. The book examines the role of extension workers in promoting best practices in animal health, nutrition, disease prevention, and welfare, as well as the adoption of new technologies and approaches in veterinary care.

The management aspect of veterinary science is given particular attention in this book. It explores the organizational structures, policies, and practices that influence veterinary clinics, animal hospitals, and animal health services. The book emphasizes the importance of effective management in ensuring the success and sustainability of veterinary practices, particularly in an era of increasing demand for animal health services. It examines the logistics of managing veterinary facilities, including staff management, resource allocation, and operational efficiency. The book also addresses the financial aspects of veterinary practice management, exploring the economic challenges faced by veterinary services, and how sound financial practices can ensure the long-term viability of these services.

The book also examines emerging trends and innovations in veterinary science, such as the increasing integration of technology into veterinary practice. Topics such as telemedicine, diagnostic advancements, and precision veterinary medicine are discussed, along with the implications of these developments for veterinary care and management. The future of veterinary services is explored, focusing on how advancements in technology and knowledge will continue to shape the veterinary field and its extension services.

Throughout the book, case studies, practical examples, and real-world applications illustrate the challenges and successes of veterinary extension and management in diverse settings. These examples provide valuable insights into the day-to-day operations of veterinary services, as well as the strategies employed to overcome obstacles and improve service delivery.

Veterinary Science: Extension and Management concludes by reflecting on the future direction of veterinary science and its management. It highlights the evolving role of veterinarians in a globalized world, where animal health issues are increasingly interconnected with human health, environmental sustainability, and agricultural productivity. The book calls for the continued professional development of veterinarians and veterinary extension workers, as well as the need for collaborative efforts between veterinary professionals, governments, NGOs, and the private sector to address the complex challenges faced by animal health systems worldwide.

This book serves as an essential resource for anyone engaged in the veterinary profession, animal care, and management. By combining theoretical knowledge with practical applications, it provides a holistic view of the veterinary field and its critical role in ensuring animal health and welfare, improving food security, and advancing public health.

Sachin P C
Veena V S
Tina B S

Contents

1

Introduction

Veterinary medicine is the branch of medicine that deals with the prevention, control, diagnosis, and treatment of disease, disorder, and injury in animals. Along with this, it deals with animal rearing, husbandry, breeding, research on nutrition, and product development. The scope of veterinary medicine is wide, covering all animal species, both domesticated and wild, with a wide range of conditions that can affect different species.

Veterinary medicine is widely practiced, both with and without professional supervision. Professional care is most often led by a veterinary physician (also known as a veterinarian, veterinary surgeon, or "vet"), but also by paraveterinary workers, such as veterinary nurses or technicians. This can be augmented by other paraprofessionals with specific specialties, such as animal physiotherapy or dentistry, and species-relevant roles such as farriers.

Veterinary science helps human health through the monitoring and control of zoonotic disease (infectious disease transmitted from nonhuman animals to humans), food safety, and through human applications via medical research. They also help to maintain food supply through livestock health monitoring and treatment, and mental health by keeping pets healthy and long-living. Veterinary scientists often collaborate with epidemiologists and other health or natural scientists, depending on type of work. Ethically, veterinarians are usually obliged to look after animal welfare. Veterinarians diagnose, treat, and help keep animals safe and healthy.

PREMODERN ERA

Archeological evidence, in the form of a cow skull upon which trepanation had been performed, shows that people were performing veterinary procedures in the Neolithic (3400–3000 BCE).

Fragments of the Kahun Papyrus on veterinary medicine, early 2nd millennium BCE). The Egyptian Papyrus of Kahun (Twelfth Dynasty of Egypt) is the first extant record of veterinary medicine.

The Shalihotra Samhita, dating from the time of Ashoka, is an early Indian veterinary treatise. The edicts of Asoka read: "Everywhere King Piyadasi (Asoka) made two kinds of medicine available, medicine for people, and medicine for animals. Where no healing herbs for people and animals were available, he ordered that they be bought and planted."

Manuscript page of Hippiatrica (14th century)

Hippiatrica is a Byzantine compilation of hippiatrics, dated to the fifth or sixthth century AD.

The first attempts to organize and regulate the practice of treating animals tended to focus on horses because of their economic significance. In the Middle Ages, farriers combined their work in horseshoeing with the more general task of "horse doctoring". The Arabic tradition of Bay?ara, or Shiyat al-Khayl, originates with the treatise of Ibn Akhi Hizam (fl. late 9th century).

In 1356, the Lord Mayor of London, concerned at the poor standard of care given to horses in the city, requested that all farriers operating within a 7-mile (11-km) radius of the City of London form a "fellowship" to regulate and improve their practices. This ultimately led to the establishment of the Worshipful Company of Farriers in 1674.

Meanwhile, Carlo Ruini's book Anatomia del Cavallo, (Anatomy of the Horse) was published in 1598. It was the first comprehensive treatise on the anatomy of a nonhuman species.

Establishment of profession

The first veterinary school was founded in Lyon, France, in 1762 by Claude Bourgelat. According to Lupton, after observing the devastation being caused by cattle plague to the French herds, Bourgelat devoted his time to seeking out a remedy. This resulted in his founding a veterinary school in Lyon in 1761, from which establishment he dispatched students to combat the disease; in a short time, the plague was stayed and the health of stock restored, through the assistance rendered to agriculture by veterinary science and art. The school received immediate international recognition in the 18th century and its pedagogical model drew on the existing fields of human medicine, natural history, and comparative anatomy.

The Odiham Agricultural Society was founded in 1783 in England to promote agriculture and industry, and played an important role in the foundation of the veterinary profession in Britain. A founding member, Thomas Burgess, began to take up the cause of animal welfare and campaign for the more humane treatment of sick animals. A 1785 society meeting resolved to "promote the study of Farriery upon rational scientific principles."

Physician James Clark wrote a treatise entitled Prevention of Disease in which he argued for the professionalization of the veterinary trade, and the establishment of veterinary colleges. This was finally achieved in 1790, through the campaigning of Granville Penn, who persuaded Frenchman Benoit Vial de St. Bel to accept the professorship of the newly established veterinary college in London. The Royal College of Veterinary Surgeons was established by royal charter in 1844. Veterinary science came of age in the late 19th century, with notable contributions from Sir John McFadyean, credited by many as having been the founder of modern veterinary research.

In the United States, the first schools were established in the early 19th century in Boston, New York City, and Philadelphia. In 1879, Iowa Agricultural College became the first land-grant college to establish a school of veterinary medicine.

VETERINARY WORKERS

Veterinary physicians

Surgery on a dog

Veterinary care and management are usually led by a veterinary physician (usually called a veterinarian, veterinary surgeon or "vet" - doctor of veterinary medicine or veterinary medical doctor). This role is the equivalent of a physician or surgeon (medical doctor) in human medicine, and involves postgraduate study and qualification.

In many countries, the local nomenclature for a vet is a protected term, meaning that people without the prerequisite qualifications and/or registration are not able to use the title, and in many cases, the activities that may be undertaken by a vet (such as animal treatment or surgery) are restricted only to those people who are registered as vet. For instance, in the United Kingdom, as in other jurisdictions, animal treatment may be performed only by registered vets (with a few designated exceptions, such as paraveterinary workers), and any persons calling themselves a vet without being registered or performing any treatment is illegal.

Most vets work in clinical settings, treating animals directly. They may be involved in a general practice, treating animals of all types; may be specialized in a specific group of animals such as companion animals, livestock, laboratory animals, zoo animals, or horses; or may specialize in a narrow medical discipline such as veterinary surgery, dermatology, cardiology, neurology, laboratory animal medicine, internal medicine, and more.

As with healthcare professionals, vets face ethical decisions about the care of their patients. Current debates within the profession include the veterinary ethics of purely cosmetic procedures on animals, such as declawing of cats, docking of tails, cropping of ears, and debarking on dogs.

A wide range of surgeries and operations is performed on various types of animals, but not all of them are carried out by vets. In a case in Iran, for instance, an eye surgeon managed to perform a successful cataract surgery on a rooster for the first time in the world.

Paraveterinary workers

Paraveterinary workers, including veterinary nurses, technicians, and assistants, either assist vets in their work, or may work within their own scope of practice, depending on skills and qualifications, including in some cases, performing minor surgery.

The role of paraveterinary workers is less homogeneous globally than that of a vet, and qualification levels, and the associated skill mix, vary widely.

ALLIED PROFESSIONS

A number of professions exist within the scope of veterinary medicine, but may not necessarily be performed by vets or veterinary nurses. This includes those performing roles which are also found in human medicine, such as practitioners dealing with musculoskeletal disorders, including osteopaths, chiropractors, and physiotherapists.

Some roles are specific to animals, but which have parallels in human society, such as animal grooming and animal massage. Some roles are specific to a species or group of animals, such as farriers, who are involved in the shoeing of horses, and in many cases have a major role to play in ensuring the medical fitness of horses.

Veterinary research

Veterinary research includes prevention, control, diagnosis, and treatment of diseases of animals, and basic biology, welfare, and care of animals. Veterinary research transcends species boundaries and includes the study of spontaneously occurring and experimentally induced models of both human and animal diseases and research at human-animal interfaces, such as food safety, wildlife and ecosystem health, zoonotic diseases, and public policy.

Clinical veterinary research

As in medicine, randomized controlled trials also are fundamental in veterinary medicine to establish the effectiveness of a treatment. Clinical veterinary research is far behind human medical research, though, with fewer randomized controlled trials, that have a lower quality and are mostly focused on research animals. Possible improvement consists in creation of networks for inclusion of private veterinary practices in randomized controlled trials.

No studies exist on the effect of community animal health services on improving household wealth and the health status of low-income farmers.

The first recorded use of regenerative stem-cell therapy to treat lesions in a wild animal occurred in 2011 in Brazil. On that occasion, the Zoo Brasília [pt] used stem cells to treat a maned wolf who had been run over by a car, which was later returned, fully recovered, to nature.

Technology in veterinary medicine

Many of the technologies used in human medicine are also used in the veterinary field, although often in slightly different ways. Veterinarians use a variety of technologies for diagnostic and therapeutic purposes to better understand and improve the health of their animal patients. Recent trends in veterinary technology have moved towards the integration of hand-held devices and consumer based technology to monitor pets and interact with veterinarians.

Veterinarian

A veterinarian (vet), also known as a veterinary surgeon or veterinary physician, is a medical professional who practices veterinary medicine. They manage a wide range of health conditions and injuries in non-human animals. Along with this, vets also play vital role in animal reproduction, animal health management, conservation, husbandry and breeding and preventive medicine like animal nutrition, vaccination and parasitic control as well as bio security and zoonotic disease surveillance and prevention.Description

In many countries, the local nomenclature for a veterinarian is a regulated and protected term, meaning that members of the public without the prerequisite qualifications and/or licensure are not able to use the title. This title is selective in order to produce the most knowledgeable veterinarians that pass these qualifications.

In many cases, the activities that may be undertaken by a veterinarian (such as treatment of illness or surgery in animals) are restricted only to those professionals who are registered as a veterinarian. For instance, in the United Kingdom, as in other jurisdictions, animal treatment may only be performed by registered veterinary physicians (with a few designated exceptions, such as paraveterinary workers), and it is illegal for any person who is not registered to call themselves a veterinarian, prescribe any drugs, or perform treatment.

Most veterinary physicians work in clinical settings, treating animals directly. These veterinarians may be involved in a general practice, treating animals of all types; they may be specialized in a specific group of animals such as companion animals, livestock, zoo animals or equines; or may specialize in a narrow medical discipline such as surgery, dermatology or internal medicine . As with other healthcare professionals, veterinarians face ethical decisions about the care of their patients. Current debates within the profession include the ethics of certain procedures believed to be purely cosmetic or unnecessary for behavioral issues, such as declawing of cats, docking of tails, cropping of ears and debarking on dogs.

Ancient Indian sage and veterinary physician Shalihotra (mythological estimate c. 2350 BCE), the son of a sage, Hayagosha, is considered the founder of veterinary sciences.

ROLES AND RESPONSIBILITIES

Veterinarians treat disease, disorder or injury in animals, which includes diagnosis, treatment and aftercare. The scope of practice, specialty and experience of the individual veterinarian will dictate exactly what interventions they perform, but most will perform surgery (of differing complexity).

Unlike in human medicine, veterinarians must rely primarily on clinical signs, as animals are unable to vocalize symptoms as a human would. In some cases, owners may be able to provide a medical history and the veterinarian can combine this information along with observations, and the results of pertinent diagnostic tests such as radiography, CT scans, MRI, blood tests, urinalysis and others.

Veterinarians must consider the appropriateness of euthanasia ("putting to sleep") if a condition is likely to leave the animal in pain or with a poor quality of life, or if treatment of a condition is likely to cause more harm to the patient than good, or if the patient is unlikely to survive any treatment regimen. Additionally, there are scenarios where euthanasia is considered due to the constrains of the client's finances.

As with human medicine, much veterinary work is concerned with prophylactic treatment, in order to prevent problems occurring in the future. Common interventions include vaccination against common animal illnesses, such as distemper or rabies, and dental prophylaxis to prevent or inhibit dental disease. This may also involve owner education so as to avoid future medical or behavioral issues.

Additionally veterinarians have important roles in public health and the prevention of zoonoses.

Employment

The majority of veterinarians are employed in private practice treating animals (75% of vets in the United States, according to the American Veterinary Medical Association).

Small animal veterinarians typically work in veterinary clinics, veterinary hospitals, or both. Large animal veterinarians often spend more time travelling to see their patients at the primary facilities which house them, such as zoos or farms.

Other employers include charities treating animals, colleges of veterinary medicine, research laboratories, animal food companies, and pharmaceutical companies. In many countries, the government may also be a major employer of veterinarians, such as the United States Department of Agriculture or the Animal and Plant Health Agency in the United Kingdom. State and local governments also employ veterinarians.

Paraveterinary worker

Paraveterinary worker is the professional of veterinary science that performs procedures autonomously or semi autonomously, as part of a veterinary assistance system. The job role varies throughout the world, and common titles include veterinary nurse, veterinary technician and veterinary assistant, and variants with the prefix of 'animal health'.

The scope of practice varies between countries, with some countries allowing suitably qualified paraveterinary workers a scope of autonomous practice, including minor surgery, whilst others restrict their workers as assistants to other professionals.

VETERINARY TECHNICIAN / NURSE

In North America, paraveterinary workers who have completed a course of study, passed an examination, and have a defined scope of practice are called veterinary technicians. Veterinary technicians hold a technician degree in Veterinary Technology. Most Canadian provinces have a formal registration process and, legally, veterinarians must hire registered veterinary technicians. In America credentialed veterinary technicians must attend a Veterinary Technician program approved by the American Veterinary Medical Association (AVMA), most of which are two year programs which confer an Associate of Science (or an Associate of Applied Science) Degree in Veterinary Technology. Graduates must pass the VTNE (Veterinary Technician National Exam) to become credentialed in their state. These credentials (whether Licensed [LVT], Registered [RVT], or Certified [CVT]) must be renewed every two years with requirements varying from state-to-state.

Human nursing associations have often claimed rights over the term "nurse". In some countries this is protected by law, and in the United States, 39 states have title protection over "Nurse", with Nursing Practice Acts mandating that only those who are Registered Nurses (RNs) may use the title "Nurse".

"Nurse" title protection was in place in the United Kingdom until 1984, where veterinary nurses were referred to as 'registered animal nursing auxiliaries', in line with the naming convention at the time for less qualified assistants in human nursing, called 'nursing auxiliaries'.

There are currently only seven countries in the world (out of ~195) who have a recognized paraveterinary profession and use the term "Veterinary Nurse" to describe those workers:

Australia

- Veterinary Nurses in only ONE STATE (Queensland). However, a VN title can be acquired with little education and only a certificate. Voluntary credentialing opened in some other states in 2018, but this is also a profession in its infancy.
- Vet Technicians, in contrast, hold a 3-year degree completed through a University program. The curriculum is not influenced by the industry or government as such and there is no quality assurance of practical skills; however, this degree has more in-depth knowledge required and exposure to much higher level skills.
- New Zealand - both Registered Veterinary Nurses and Veterinary Technicians
- Veterinary Technicians have the higher degree (3 years of school as opposed to 2)
- South Africa - Veterinary Nurse and Animal Health Technician
- Each title conferred has completely separate curricula, with the AHT being more like a Nurse Practitioner (in the US); aka a mid-level medical professional.
- Veterinary Nurses in SA are more like veterinary assistants in the US
- Denmark - title translates to Veterinary Nurse
- Norway - title translates to Veterinary Nurse

There is an effort to change the title of credentialed Veterinary Technicians in the United States but legislative efforts have failed in all four states where the name change has been introduced to date. The VNI has spent over $200,000 with no clear accounting of where the money has been spent and no success. Many state Veterinary Technician associations and Veterinary Medical Associations oppose the effort to change the title of the profession in the United States, and the AVMA has refused to support the name change (but does support the other stated goals of the VNI).

Veterinary assistant

In most countries, a veterinary assistant is a person with fewer or no formal animal health qualifications, who has no autonomous practice, but who is designated to assist other professionals.

Training programs are often workplace-based, and no formal licence or certification is required to perform the role.

In the US, veterinary assistants have the option to earn a certificate of completion by taking basic animal health classes about contagious diseases, animal restraint, record keeping, work place safety, administration, etc. Having the knowledge of the basics allows for the development of trust between the veterinarian and the assistant, as well as smoother job training of veterinary assistants. Their scope of practice remains limited and equal to many on the job trained staff. Local laws restrict what activities a veterinary assistant may perform, as some procedures may only be legally completed by a licensed veterinary technician, including IV anesthesia induction, oral surgery, splinting and casting, and in some states, administering the rabies vaccine.

History

Veterinarians have had assistance from staff throughout their existence of the profession, but the first organised paraveterinary workers were the canine nurses trained by the Canine Nurses Institute in 1908, and announced in the magazine 'The Veterinary Student'. According to the founder, they would "carry out directions of the veterinary surgeon, meet a genuine need on the part of the dog owners, and at the same time provide a reasonably paid occupation for young women with a real liking for animals".

In 1913, the Ruislip Dog Sanatorium was founded, and employed nurses to care for unwell dogs and in the 1920s, at least one veterinary surgery in Mayfair employed qualified human nurses to tend the animals. In the mid-1930s, the early veterinary nurses approached the Royal College of Veterinary Surgeons for official recognition, and in 1938 the Royal Veterinary College had a head nurse appointed, but the official recognition was not given until 1957, first as veterinary nurses, but changed within a year to Royal Animal Nursing Auxiliaries (RANAs) following objection from the human nursing profession.

In 1951, the first formal paraveterinary role was created by the United States Air Force who introduced veterinary technicians, and this was followed in 1961 by a civilian programme at the State University of New York (SUNY) Agricultural and Technical College. In 1965 Walter Collins, DVM received federal funding to develop model curricula for training technicians. He produced several guides over the next seven years, and for this work he is considered the "father of veterinary technology" in the United States.

In 1984, the term veterinary nurse was formally restored to paraveterinary workers in the United Kingdom.

Role and responsibilities

Veterinary Technician Code of Ethics

1. Veterinary technicians shall aid society and animals by providing excellent care and services for animals.
2. Veterinary technicians shall prevent and relieve the suffering of animals with competence and compassion.
3. Veterinary technicians shall remain competent through commitment to life-long learning.
4. Veterinary technicians shall promote public health by assisting with the control of zoonotic diseases and educating the public about these diseases.
5. Veterinary technicians shall collaborate with other members of the veterinary medical profession in efforts to ensure quality health care services for all animals.
6. Veterinary technicians shall protect confidential information provided by clients, unless required by law or to protect public health.
7. Veterinary technicians shall assume accountability for individual professional actions and judgments.
8. Veterinary technicians shall safeguard the public and the profession against individuals deficient in professional competence or ethics.
9. Veterinary technicians shall assist with efforts to ensure conditions of employment consistent with the excellent care for animals.
10. Veterinary technicians shall uphold the laws/regulations that apply to the technician's responsibilities as a member of the animal health care team.

11. Veterinary technicians shall represent their credentials or identify themselves with specialty organizations only if the designation has been awarded or earned.

The scope of practice for paraveterinary workers varies by jurisdiction, and by qualification level. In some places, more than one grade of paraveterinary worker exists. For instance, in the United Kingdom there are both veterinary nurses, who are qualified professionals with a protected title, and veterinary assistants, who do not have a single level of qualification which they must attain, and whose title is not protected. Furthermore, job roles may be divided further into roles such as Veterinary Surgical Technician, Veterinary Emergency and Critical Care Technician, Veterinary Technician Anesthesia Specialist, etc.

In the United States, veterinary technicians can become registered both nationally and by their state of residence by taking board exams. Not all states require a state test. States that license veterinary technicians include Alaska, Alabama, Delaware, Georgia, Michigan, Nebraska, Nevada, New York, North Dakota, Tennessee, Texas, Virginia, and Washington. Once they are registered, they are allowed to perform certain tasks with or without the supervision of another professional. To see a list by state, click here to view some of these tasks. In order to keep their skills and knowledge up to par, some states require registered technicians to engage in continued education before renewing their license.

At the higher levels, veterinary nurses or technicians may be able to practice skills autonomously, including examinations and minor surgery on animals, without the direct supervision of other professionals.

Paraveterinary workers are likely to assist other professionals, or perform by themselves, medical skills such as observations (e.g. taking and recording pulse, temperature, respiration etc.), wound and trauma management (e.g. cleaning and dressing wounds, applying splints etc.), physical interventions (e.g. catheterizations, ear flushes and venipuncture) and preparing and analysing biological samples (e.g. performing skin scrapings, microbiology, urinalysis, and microscopy).

Dependent on their scope of practice and training, they may also be called upon to operate diagnostic screening equipment, including electrocardiographic, radiographic and ultrasonographic instruments, including complex machines such as computed tomography, magnetic resonance imagers and gamma cameras. In veterinary hospitals, veterinary technicians can perform complete blood counts, differential counts, and morphologic examinations of blood.

Veterinary technicians commonly assist other professionals in surgery by providing correct equipment and instruments and by assuring that monitoring and support equipment are in good working condition. They may also maintain treatment records and inventory of all pharmaceuticals, equipment and supplies, and help with other administrative tasks within a veterinary practice, such as client education.

Client education plays a key role of the veterinary technician's responsibilities, such as effectively communicatIng complex medical instructions in a positive and understandable way, and to facilitate the patient's care as an intermediary between the doctor, hospital and the patient. In this way, open lines of communication are established that can benefit the patient and hospital.

Education and qualification

The level of education of a paraveterinary worker will depend on the role they are performing, and the veterinary medico-legal framework for the area in which they are working. Many areas employ veterinary assistants, who have a simple role to directly assist other professionals, and may hold no formal qualification or training, or have been trained on the job.

Higher level paraveterinary workers, such as veterinary nurses or veterinary technicians, who have a scope of autonomous practice which they are expected to perform without instruction, are likely to have both formal qualifications and in many jurisdictions will also require a formal registration with a monitoring body.

In countries where the role of veterinary workers is most advanced, the qualification required is likely to be based in higher education, such as in the United States or Canada where veterinary technicians must normally gain an associate degree at an institution recognised by the American Veterinary Medical Association or Canadian Veterinary Medical Association, and can choose to study for an extended period to gain a bachelor's degree (which in America may confer the title 'technologist', rather than 'technician'), or the United Kingdom, where veterinary nurses enter the profession through either a two-year diploma programmer or through completion of a foundation degree or honours degree.

In almost all cases, regardless of the level of formalized training, a high level of practical experience is usually required prior to a student being fully qualified, which may be completed as part of their course, or during a post-qualification period. This may require maintenance of a log of all work completed, which may need to be signed by a supervising professional (such as the vet or senior member of the veterinary staff) to indicate competence. In some cases, such as in the United States, video records may be required of some procedures, which may then be examined by the awarding or registration body.

Many countries, including the United Kingdom, Canada, and parts of the United States, restrict some elements of practice, and may restrict use of the recognised name, to those people currently registered with an appropriate licensing body, meaning that it would be illegal for any person not on the register to represent themselves as a paraveterinary worker, or to perform some of the procedures that a licensed professional could. The precise details of these restrictions vary widely between legal areas, and neighbouring areas may have different policies, as is the case in the various states of the US.

This licensing body may have its own requirements for maintaining a registration, and those who hold the requisite academic qualification may still have to complete a further range of exams or tests to become registered. For instance, in the United States, most areas use the Veterinary Technician National Exam, and this will be used by the state licensing authority (such as a state veterinary medical association) to qualify an applicant to become a registered veterinary technician.

In some cases, those people who qualified before the introduction of formal academic qualification requirements may still be working as paraveterinary workers, and may still be entered on a required register through the use of grandfather rights. For instance, in some states of the US, people with a set number of years or hours of experience assisting other professionals could sit for the Veterinary Technician

National Exam, however this route was phased out in 2011, and future candidates must have an academic qualification.

Specialty certification

Beyond credentialing as a veterinary technician specialty certification is also available to technicians with advanced skills. To date there are specialty recognitions in: emergency & critical care, anesthesiology, dentistry, small animal internal medicine, large animal internal medicine, cardiology, oncology, neurology, zoological medicine, equine veterinary nursing, surgery, behavior, nutrition, clinical practice (canine/feline, exotic companion animal, and production animal sub-specialties) and clinical pathology. Veterinary Technician Specialists carry the additional post-nominal letters "VTS" with their particular specialties indicated in parentheses. As veterinary technology evolves more specialty academy recognitions are anticipated.

The need for veterinary assistants will continue to rise as pet ownership in the United States is growing. Responsible pet ownership and animal safety becoming more significant's contributing to the increasing demand. According to the U.S. Department of Labor, the number of veterinary assistant jobs is expected to grow up to 19% by 2026, much faster than the average career.

ANIMAL TECHNOLOGY

Animal technology refers to the practices of keeping, breeding and providing care for animals that are used for scientific purposes, such as captive in a laboratory. Animal technology is one of the recognized professional areas held by registered science technicians, and animal technologists, also called animal technicians, play a key role in supporting biomedical research by ensuring that animals are available for study. Principal areas of animal technology include animal husbandry and breeding, providing day-to-day care for laboratory animals, ensuring compliance with animal welfare practices and legal issues and performing essential scientific procedures. Technical qualifications for animal technologists and the regulations they must adhere to vary by country, but in many parts of the world animal technology is a highly structured profession that forms part of laboratory management. Animal technology is related to the field of animal management and technologists are often specialize in working with particular species of animals, either in the laboratory or in the field.

Veterinary medicine in the United Kingdom

Veterinary medicine in the United Kingdom is the performance of veterinary medicine by licensed professionals. It is strictly regulated by the statute law, notably the Veterinary Surgeons Act 1966. Veterinary medicine is led by veterinary physicians, termed "veterinary surgeons" (with a different meaning to how it is used in some other anglophone countries, where it denotes a surgical specialist), normally referred to as "vets".

Vets are often assisted by registered veterinary nurses, who are able to both assist the vet and to autonomously practise a range of skills of their own, including minor surgery under direction from a responsible vet.

Other professionals are also permitted to perform some animal treatment, through exemptions in the law, and these include manipulation techniques such as physiotherapy, chiropractic and osteopathy. Other alternative medicine therapies, such as homeopathy, acupuncture, phytotherapy and aromatherapy may only be performed by a licensed veterinary surgeon.

The practice of veterinary medicine in the United Kingdom is regulated by the Royal College of Veterinary Surgeons (RCVS), which licenses both veterinary surgeons and veterinary nurses.

Veterinary surgeons

In order to practise veterinary medicine in the UK, a veterinary surgeon must hold a current registration with the RCVS. This requires a qualifying degree, usually in veterinary science or veterinary medicine. From March 2015, veterinary surgeons registered with the RCVS in the UK may optionally use the courtesy title Dr. This makes it the third clinical degree in the UK, after medicine and dentistry, to allow the use of the title Dr. The origins of veterinary surgeons parallel to human surgery are reflected in human medicine where qualified surgeons also drop their Dr designation and revert to their original title.

Education

Veterinary medicine degree courses are usually five years in length, holders of a science degree may take a four-year accelerated course and all intercalated degrees take six years. There are a limited number of places on veterinary courses each year, with only eight UK universities offering the degree (Bristol, Cambridge, Edinburgh, Glasgow, Liverpool, Nottingham, the Royal Veterinary College in London, and Surrey). The University of Central Lancashire (UCLan) also has a Veterinary School and will offer the veterinary medicine degree from September 2023.

Continuing professional development (CPD) is a mandatory and key part of career development. The RCVS recommends a minimum of 105 hours CPD over a three-year period. The RCVS Professional Development Phase (PDP) that was launched for new graduates in 2007 provides a structured approach to guide the new graduate towards the professional competences they need to develop in either small animal, equine or production animal practice.

Specialism

Vets may choose to specialise in various areas of veterinary medicine, through certificate qualifications, modular certificates or diplomas, with each speciality taking around two years to complete. Certificates cover a wide range of areas, including small animal medicine, small animal surgery, large animal medicine, welfare ethics and law, public health, cardiology, orthopaedics and advanced veterinary practice (Cert AVP).

There is a large and potentially confusing array of post-nominal titles in the UK veterinary profession. Levels of credibility vary. In particular, Certificate level qualification does not qualify a veterinary surgeon as a specialist. With further training, extensive professional experience and by publishing articles in a particular subject area, it is possible to gain Royal College of Veterinary Surgeons (RCVS) Recognised Specialist Status. In 2012, the ruling council of the RCVS adopted a report from the Calman Committee into specialisation in the UK veterinary profession, accepting that only those veterinary surgeons recognised by the RCVS as specialists and placed on a register held by the RCVS could be truly held up as "Specialist".

Vets may undertake the training to become an Official Veterinarian (OV) which authorises them to carry out tasks on behalf of the Department for Environment, Food and Rural Affairs, such as testing cattle for tuberculosis or issuing of documentation for the export of animals and animal products.

Employment

Most veterinary surgeons work in private practice, either in a general practice, or specialising in one type of animal (small animal, equine, zoo animal etc.). Newly qualified veterinary surgeons usually work as assistants for some time before being offered the opportunity to become a partner or a principal. Becoming a partner involves increased responsibility, the need for more business and management skills and a financial input into the practice.

Some vets are also employed by animal welfare charities who offer treatment to the public, such as Royal Society for the Prevention of Cruelty to Animals (RSPCA), the People's Dispensary for Sick Animals (PDSA) and The Blue Cross.

There are also opportunities to work for government services, including APHA (the Animal and Plant Health Agency) who are responsible for the control and eradication of major notifiable diseases, animal welfare, promotion of international trade and certain public health functions related to residues in meat and investigation of food safety incidents, and scanning surveillance, or the Veterinary Medicines Directorate (VMD) who license veterinary medicines.

It is also possible to pursue a research and/or teaching career within universities or research bodies.

Veterinary nurses

Veterinary Nurses are the primary paraveterinary workers in the United Kingdom and work alongside vets. Veterinary Nurses must be registered and follow a strict code of conduct. Veterinary Nurses have a scope of autonomous practice within which they can act for the animals they treat. Under schedule 3 of the Veterinary Surgeons Act they can perform many complex procedures include minor surgery, admission of intravenous fluid therapy and parenteral nutrition, performing diagnostic imaging and monitoring anaesthesia.

Preventative medicine is also an important part of the veterinary nurse's role with nurse clinics and consultations becoming increasingly common. Nurse led clinics may cover such areas as nutrition and weight management, management of the diabetic or senior patient, parasite control, vaccinations, puppy and kitten socialisation, dental care and wound care.

Registration

All veterinary nurses must be registered with the Royal College of Veterinary Surgeons (RCVS). Registered Veterinary Nurses have dispensations in law (the Veterinary Surgeons Act 1966, amended in 2002) to undertake certain procedures to include minor surgery and anaesthesia on animals under veterinary direction.

Registered Veterinary Nurses (RVNs) are bound by a code of professional conduct and are obliged to maintain their professional knowledge and skills through ongoing CPD. Those VN's listed after 2002 are automatically registered.

Education

RVNs train for the Register through either a two to three year further education diploma programme or via a qualifying foundation or honours degree which can be three to four years. All student nurses must complete a minimum of 2100 hours of work experience in general veterinary practice during their training.

RCVS badges are engraved with the nurse's personal badge number, and as of 2012 are available with the additional engraving 'registered' for RVNs.

VN's can further their formal training and gain additional postnominal letters by completing one of many specialist certificates and/or by achieving the RCVS Diploma in Advanced Veterinary Nursing (DipAVN) by following a course of study in one or more of three pathways: Small animal nursing, Equine nursing, and Veterinary nursing education. The streamlined RCVS DipAVN replaced the DipAVN(Medical) and DipAVN(Surgical) formerly administered by the BVNA up until 2005. The DipAVN is offered through Myerscough College in Lancashire since 2007 and more recently through Harper Adams University, some modules are also available from the Royal Veterinary College in conjunction with the University of London.

Animal nursing assistants

Lay staff, without formal qualification or status may be called animal nursing assistants (the BVNA approved term), veterinary care assistants, or veterinary care assistants. These lay staff have usually undertaken some basic veterinary nurse training but are limited by law as to the procedures they may undertake on animals. They work alongside qualified vets and veterinary nurses to provide care and support to animal patients and their owners.

2

Study of Animals Life Cycles

Invertebrate animals have a rich variety of life cycles, especially among those forms that undergo metamorphosis, a radical physical change. Butterflies, for instance, have a caterpillar stage (larva), a dormant chrysalis stage (pupa), and an adult stage (imago). One remarkable aspect of this development is that, during the transition from caterpillar to adult, most of the caterpillar tissue disintegrates and is used as food, thereby providing energy for the next stage of development, which begins when certain small structures (imaginal disks) in the larva start growing into the adult form. Thus, the butterfly undergoes essentially two periods of growth and development (larva and pupa–adult) and two periods of small size (fertilized egg and imaginal disks). A somewhat similar phenomenon is found in sea urchins; the larva, which is called a pluteus, has a small, wartlike bud that grows into the adult while the pluteus tissue disintegrates. In both examples it is as if the organism has two life histories, one built on the ruins of another.

Another life-cycle pattern found among certain invertebrates illustrates the principle that major differences between organisms are not always found in the physical appearance of the adult but in differences of the whole life history. In the coelenterate *Obelia*, for example, the egg develops into a colonial hydroid consisting of a series of branching *Hydra*-like organisms called polyps. Certain of these polyps become specialized (reproductive polyps) and bud off from the colony as free-swimming jellyfish (medusae) that bear eggs and sperm. As with caterpillars and sea urchins, two distinct phases occur in the life cycle of *Obelia*: the sessile (anchored), branched polyps and the motile medusae. In some related coelenterates the medusa form has been totally lost, leaving only the polyp stage to bear eggs and sperm directly. In still other coelenterates the polyp stage has been lost, and the medusae produce other medusae directly, without the sessile stage. There are, furthermore, intermediate forms between the extremes.

NATURAL SELECTION AND REPRODUCTION

The significance of biological reproduction can be explained entirely by natural selection (see evolution: The concept of natural selection). In formulating his theory of natural selection, Charles Darwin realized that, in order for evolution to occur, not only must living organisms be able to reproduce themselves but the copies must not all be identical; that is, they must show some variation. In this way the more successful variants would make a greater contribution to subsequent generations in the number of offspring. For such selection to act continuously in successive generations, Darwin also recognized that the variations had to be inherited, although he failed to fathom the mechanism of heredity. Moreover, the amount of variation is particularly important. According to what has been called the principle of compromise, which itself has been shaped by natural selection, there must not be too little or too much variation: too little produces no change; too much scrambles the benefit of any particular combination of inherited traits.

Of the numerous mechanisms for controlling variation, all of which involve a combination of checks and balances that work together, the most successful is that found in the large majority of all plants and animals—*i.e.,* sexual reproduction. During the evolution of reproduction and variation, which are the two basic properties of organisms that not only are required for natural selection but are also subject to it, sexual reproduction has become ideally adapted to produce the right amount of variation and to allow new combinations of traits to be rapidly incorporated into an individual.

The evolution of reproduction

An examination of the way in which organisms have changed since their initial unicellular condition in primeval times shows an increase in multicellularity and therefore an increase in the size of both plants and animals. After cell reproduction evolved into multicellular growth, the multicellular organism evolved a means of reproducing itself that is best described as life-cycle reproduction. Size increase has been accompanied by many mechanical requirements that have necessitated a selection for increased efficiency; the result has been a great increase in the complexity of organisms. In terms of reproduction this means a great increase in the permutations of cell reproduction during the process of evolutionary development.

Size increase also means a longer life cycle, and with it a great diversityof patterns at different stages of the cycle. This is because each part of the life cycle is adaptive in that, through natural selection, certain characteristics have evolved for each stage that enable the organism to survive. The most extreme examples are those forms with two or more separate phases of their life cycle separated by a metamorphosis, as in caterpillars and butterflies; these phases may be shortened or extended by natural selection, as has occurred in different species of coelenterates.

To reproduce efficiently in order to contribute effectively to subsequent generations is another factor that has evolved through natural selection. For instance, an organism can produce vast quantities of eggs of which, possibly by neglect, only a small percent will survive. On the other hand, an organism can produce very few or perhaps one egg, which, as it develops, will be cared for, thereby greatly increasing its chances for survival. These are two strategies of reproduction; each has its advantages and disadvantages. Many other considerations of the natural history and structure of the organism determine, through natural selection, the strategy that is best for a particular species; one of these is that any species must not produce too few offspring (for it will become extinct) or too many (for it may also become extinct by overpopulation and disease). The numbers of some organisms fluctuate cyclically but always remain between upper and lower limits. The question of how, through natural selection, numbers of individuals are controlled is a matter of great interest; clearly, it involves factors that influence the rate of reproduction.

The evolution of variation control

Because inherited variation is largely handled by genes in the chromosomes, organisms that reproduce sexually require a single-cell stage in their life cycle, during which the haploid gamete of each parent can combine to form the diploid zygote. This is also often true in organisms that reproduce asexually, but in this case the asexual reproductive bodies (*e.g.*, spores) are small and hence are effectively dispersed.

The amount of variation is controlled in a large number of ways, all of which involve a carefully balanced set of factors. These factors include whether the organism reproduces asexually or sexually; the mutation(gene change) rate; the number of chromosomes; the amount of exchange of parts of chromosomes (crossing over); the size of the individual (which correlates with complexity and generation time); the size of the population; the degree of inbreeding versus outbreeding; and the relative amounts and position of haploidy and diploidy in the life cycle. It is clear, therefore, that the mode of reproduction influences the amount of variation and vice versa; the two together permit natural selection to operate, and selection in turn modifies the mechanisms of reproduction and variation.

Animal reproductive system

The role of reproduction is to provide for the continued existence of a species; it is the process by which living organisms duplicate themselves. Animals compete with other individuals in the environmentto maintain themselves for a period of time sufficient to enable them to produce tissue nonessential to their own survival, but indispensable to the maintenance of the species. The additional tissue, reproductive tissue, usually becomes separated from the individual to form a new, independent organism.

This article describes the reproductive systems in metazoans (multicelled animals) from sponges to mammals, exclusive of humans. It focuses on the gonads (sex organs), associated ducts and glands, and adaptations that aid in the union of gametes—*i.e.*, reproductive cells, male or female, that are capable of producing a new individual by union with a gamete of the opposite sex. Brief mention is made of how the organism provides for the development of embryos and of the regulatory role of gonads in vertebrate cycles. For discussion of reproduction in humans, see reproductive system, human.

Unlike most other organ systems, the reproductive systems of higher animals have not generally become more complex than those of lower forms. Asexual reproduction (*i.e.*, reproduction not involving the union of gametes), however, occurs only in the invertebrates, in which it is common, occurring in animals as highly evolved as the sea squirts, which are closely related to the vertebrates. Temporary gonads are common among lower animals; in higher animals, however, gonads are permanent organs. Hermaphroditism, in which one individual contains functional reproductive organs of both sexes, is common among lower invertebrates; yet separate sexes occur in such primitive animals as sponges, and hermaphroditism occurs in animals more highly evolved—*e.g.*, the lower fishes. Gonads located on or near the animal surface are common in the lowest invertebrates, but in higher animals they tend to be more deeply situated and often involve intricate duct systems. In echinoderms, which are among the highest invertebrates, the gonads hang directly into the sea and spill their gametes into the water. In protochordates, gametes are released into a stream of respiratory water that passes directly into the sea. Duct systems of the invertebrate flatworms (Platyhelminthes) are relatively complex, and those of specialized arthropods (*e.g.*, insects, spiders, crabs) are more complex than those of any vertebrate. Copulatory organs occur in flatworms, but copulatory organs are not ubiquitous among vertebrates other than reptiles and mammals. The trend toward fewer eggs and increased parental care in higher animals may account for the relative lack of complexity in the reproductive systems of some advanced forms. Whereas trends toward increasing structural complexity have often been reversed during evolution, reproductive behaviour patterns in many phylogenetic (*i.e.*, evolutionary) lines have become more complicated in order to enhance the opportunity for fertilization of eggs and maximum survival of offspring.

A direct relationship exists between behaviour and the functional state of gonads. Reproductive behaviour induced principally but not exclusively by organic substances called hormones promotes the union of sperm (spermatozoa) and eggs, as well as any parental care accorded the young. There are a number of reasons why behaviour must be synchronized with gonadal activity. Chief among these are the following:

Individuals of a species must congregate at the time the gonads contain mature gametes. This often entails migration, and some members of all major vertebrate groups migrate long distances to gather at spawning grounds or rookeries.

Individuals with gametes ready to be shed must recognize members of the opposite sex. Recognition is sometimes by external appearance or by chemical substances (pheromones), but sex-linked behaviour is often the only signal.

Geographical territories frequently must be established and aggressively defended.

The building of nests, however simple, is essential reproductive behaviour in many species.

When fertilization of aquatic forms is external, sperm and eggs must be discharged at approximately the same time into the water, since gametes may be quickly dispersed by currents. Courtship, often involving highly intricate behaviour patterns, serves to release the gametes of both mating individuals simultaneously.

When fertilization is internal, willingness of the female to mate is often essential. Female mammals not in a state of willingness to mate not only will not mate but may injure or even kill an aggressive male. The unwillingness of a female mammal to mate when mature eggs are not present prevents loss of sperm needed to preserve the species.

Parental care of fertilized eggs by one parent or the other has evolved in many species. Parental behaviour includes fanning the water or air around the eggs, thereby maintaining appropriate temperature and oxygen levels; secretion of oxygen from a parent's gills; transport of eggs on or in the parental body (including the mouth of some male parents); and brooding, or incubation, of eggs.

Some species extend parental care into the postnatal period, feeding and protecting the offspring. Such behaviour patterns are adaptations for survival and thus are essential; all are induced by the nervous and endocrine systems and are typically cyclical, because gonadal activity is cyclical.

REPRODUCTIVE SYSTEMS OF INVERTEBRATES

Gonads, associated structures, and products

Although asexual reproduction occurs in many invertebrate species, most reproduce sexually. The basic unit of sexual reproduction is a gamete (sperm or egg), produced by specialized tissues or organs called gonads. Sexual reproduction does not necessarily imply copulation or even a union of gametes. As might be expected of such a large and diverse group as the invertebrates, many variations have evolved to ensure survival of species. In many lower invertebrates, gonads are temporary organs; in higher forms, however, they are permanent. Some invertebrates have coexistent female and male gonads; in others the same gonad produces both sperm and eggs. Animals in which both sperm and eggs are produced by the same individual (hermaphroditism) are termed monoecious. In dioecious species, the sexes are separate. Generally, the male gonads ripen first in hermaphroditic animals (protandry); this tends to ensure cross-fertilization. Self-fertilization is normal, however, in many species, and some species undergo sex reversal.

Sponges, coelenterates, flatworms, and aschelminths

Sponges are at a cellular level of organization and thus do not have organs or even well-developed tissues; nevertheless, they produce sperm and eggs and also reproduce asexually. Some species of sponge are monoecious, others are dioecious. Sperm and eggs are formed by aggregations of cells called amoebocytes in the body wall; these are not considered gonads because of their origin and transitory nature.

In hydrozoan coelenterates, temporary gonads are formed by groups of cells in either the epidermis (outer cell layer) or gastrodermis (gut lining), depending on the species; scyphozoan and anthozoan coelenterates generally have gonads in the gastrodermis. The origin and development of gonads in coelenterates, particularly freshwater species, are often associated with the seasons. Freshwater hydrozoans, for example, reproduce asexually until the onset of cold weather, which stimulates them to form testes and ovaries. Colonial hydrozoans asexually produce individuals known as polyps. Polyps, in turn, give rise to free swimming stages (medusae), in which gonads develop. The body organization of sponges and coelenterates is such that most of their cells are in intimate contact with the environment; consequently, gametes are shed into the water, and no ducts are necessary to convey them to the outside.

In contrast to sponges and coelenterates, platyhelminths generally have well-developed organ systems of a permanent nature and, in addition, have evolved secondary reproductive structures to convey sex products. One exception is the acoels, a group of primitive turbellarians; they lack permanent gonads, and germinal cells develop from amoebocytes in much the same manner as in sponges. The majority of flatworms, however, are monoecious, the primary sex organs consisting of one or more ovaries and testes. The tube from the ovary to the outside is called the oviduct; it often has an outpocketing (seminal receptacle) for the storage of sperm received during copulation. In many species the oviduct receives a duct from yolk (vitelline) glands, whose cells nourish the fertilized egg. Beyond the entrance of the duct from the yolk glands the oviduct may be modified to secrete a protective capsule around the egg before it is discharged to the outside. The male organs consist of testes, from which extend numerous tubules (vasa efferentia) that unite to form a sperm duct (vas deferens); the latter becomes an ejaculatory duct through which sperm are released to the outside. The sperm duct may exhibit expanded areas that store sperm (seminal vesicles), and it may be surrounded by prostatic cells that contribute to the seminal fluid. The sperm duct eventually passes through a copulatory organ. The same basic structural pattern, somewhat modified, is found in most higher invertebrates.

Aschelminthes (roundworms) are mostly dioecious; frequently there are external differences between males and females (sexual dimorphism). The males are generally smaller and often have copulatory spicules. Nematodes have relatively simple reproductive organs, a tubular testis or ovary being located at the end of a twisted tube. The portion of the female tract nearest the ovary forms a uterus for temporary storage of fertilized eggs. Some species lay eggs, but others retain the egg in the uterus until the larva hatches. The sperm are released into a cavity called the cloaca. A number of free-living nematodes are capable of sex reversal—if the sex ratio in a given population is not optimal or if environmental conditions are not ideal, the ratio of males to females can be altered. This sometimes results in intersexes; i.e., females with some male characteristics. Hermaphroditism occurs in nematodes, and self-fertilization in such species is common.

ANNELIDSAND MOLLUSKS

Annelids have a well-developed body cavity (coelom), a part of the lining of which gives rise to gonads. In some annelids, gonads occur in several successive body segments. This is true, for example, in polychaetes, most of which are dioecious. Testes and ovaries usually develop, though not invariably, in many body segments; and the sperm and eggs, often in enormous numbers, are stored in the coelom. Fertilization is external. In oligochaetes (all of which are monoecious) on the other hand, the gonads develop in a few specific segments. Sperm are stored in a seminal vesicle and eggs in an egg sac, rather than in the coelom. A portion of the peritoneum, the membrane lining the coelom, becomes a saclike seminal receptacle that stores sperm received from the mate. The earthworm, Lumbricus terrestris, is an example of a specialized annelid reproductive system. Female organs consist of a pair of ovaries in segment 13; a pair of oviducts that open via a ciliated funnel (i.e., with hairlike structures) into segment 13 but open to the exterior in segment 14; an egg sac near each funnel; and a pair of seminal receptacles in segment 9 and another in segment 10. Male organs consist of two pairs of minute testes in segments 10 and 11, each associated with a ciliated sperm funnel leading to a tiny duct, the vas efferens. The two ducts on each side lead to a vas deferens that opens in segment 15. Testes and funnels are contained within two of three pairs of large seminal vesicles that occupy six body segments. Leeches (Hirudinea), also monoecious, have one pair of ovaries and a segmentally arranged series of testes with duct systems basically similar to those of earthworms.

Although mollusks have a close evolutionary kinship to annelids, they have reduced or lost many structures characteristic of segmented worms. The coelom persists only as three regional cavities: gonadal, nephridial (kidney), and pericardial (heart). In ancestral forms these were interconnected so that gametes from the gonad passed through the pericardial cavity, the nephridial cavity, then to the outside through a nephridial pore. The various groups of mollusks have tended to modify this arrangement, with the result that gonads have their own pore; among amphineurans, for example, the sexes are usually separate, and there is one gonad with an associated pore. Gastropods show considerable variability, but generally one gonad (ovary, testis, or ovotestis—a structure combining the functional gonads of both sexes) is located in the visceral hump and connected to the outside by a remnant of the right kidney. In hermaphroditic forms, one duct carries sperm as well as eggs. The gonadal ducts of gastropod females often secrete a protective capsule around the fertilized eggs; in males, the terminal portion of the duct is sometimes contained in a copulatory organ. Pelecypods may be either monoecious or dioecious, but the gonads are usually paired. In mussels and oysters, the gonads open through the nephridial pore, but in clams the reproductive system opens independently. The cephalopods are all dioecious. The single testis or ovary releases its products into the pericardial cavity and this, in turn, leads to a gonopore, the external opening. The oviduct of the squidis terminally modified to form a shell gland. The male system is more complex—the gonoduct leads into a seminal vesicle where a complicated torpedo-shaped sperm case (spermatophore) is secreted and contains the sperm. Spermatophores are then stored in a special structure (Needham's sac) until copulation occurs.

A remarkable characteristic of some mollusks is the ability to alter their sex. Some species are clearly dioecious; however, among the monoecious species there is considerable variability in their hermaphroditic condition. In some species, male and female gonads, although in the same individual, are independent functionally and structurally. In others, an ovotestis produces both sperm and eggs. Oysters display a third condition; young oysters have a tendency toward maleness, but, if water temperature or food availability is altered, some individuals develop into females. Later, a reversal to the male condition may occur. The sexual makeup of an entire oyster population also has a seasonal aspect; in harmony with the group, an individual may undergo several alterations in the course of a year. A similar phenomenon, called consecutive sexuality, occurs in limpets. These gastropods stack themselves in piles, with the younger animals on top. The animals on top are males with well-developed testes and copulatory organs; those in the middle are hermaphroditic; those on the bottom are females, having lost the testes and copulatory organ (penis) by degeneration. A decrease in the number of females in a stack induces males to assume female characteristics, but the transition is retarded when an excess of females is present. The degree of maleness or femaleness is probably controlled in part by environmental and internal factors.

Form and Function

To stay alive, grow, and reproduce, an animal must find food, water, and oxygen, and it must eliminate the waste products of metabolism. The organ systems typical of all but the simplest of animals range from those highly specialized for one function to those participating in many. The more basic functional systems are treated below from a broadly comparative basis.

SUPPORT AND MOVEMENT

A skeleton can support an animal, act as an antagonist to musclecontraction, or, most commonly, do both. Because muscles can only contract, they require some other structure to stretch them to their noncontracted (relaxed) state. Another set of muscles or the skeleton itself can act as an antagonist to muscle contraction. Only elastic skeletons can act without an antagonist; all antagonistic muscles act through a skeleton, which can be either rigid, flexible, or hydrostatic.

Types of skeletons and their distribution

Hydrostatic skeletons are the most prevalent skeletal system used by animals for movement and support. A minimal hydroskeleton resembles a closed container. The walls are two layers of muscles (antagonists) oriented at right angles to one another; the inside contains an incompressible fluid or gel. The contraction of one set of muscles exerts a pressure on the fluid, which is forced to move at right angles to the squeezing antagonist. The movement of the fluid stretches the other set of muscles, which can then contract to stretch its antagonist back to its relaxed position. The net result is an alternating change in the shape of the container. Locomotion as varied as crawling, burrowing, somersaulting, looping, or even walking is possible when the container has some means of traction against a substrate: the system extends forward from the point of attachment, attaches at a more forward point, releases posteriorly, and contracts forward. Hydroskeletons are also important in nonlocomotory muscular systems, such as hearts or intestines, which move blood or food, respectively. Contraction-relaxation cycles push in one direction only when the system has structures that prevent backflow.

Hydroskeletons become less efficient when fluid is lost. The optimal volume of fluid for a particular system must remain constant for effective contraction and expansion of the antagonistic muscles. If too much fluid is lost, the animal becomes limp and neither muscle can stretch; when too much fluid is gained, the animal becomes bloated and neither muscle can contract. Those coelenterates that use a hydroskeleton regularly face a loss of pressure because their skeleton is also their gut. Freshwater animals tend to become bloated as water diffuses into their salty cells, but terrestrial animals with hydroskeletons tend to become limp as they dry. Solutions to water loss tend to be partial because impermeable barriers, such as a shell, tend not to be very flexible, thus negating the use of a hydroskeleton for movement. Terrestrial animals with locomotory hydroskeletons (e.g., snails and earthworms) are restricted in their activity to moist conditions.

Partitioning a hydroskeleton into many small, separate, but coordinated units facilitates locomotion. In an earthworm, for example, a front group of segments narrows together, thereby elongating that part of the worm. If there were no partitions between the segments, the fluid would flow farther back, providing little elongation. Widened segments behind these initial segments anchor the worm, and its head moves forward. The process then reverses in a wave, and the posterior end moves forward. Metamerism, or the partitioning of the coelom, is thought to have evolved in ancestral annelids to improve their ability as burrowers in the bottom mud of the ocean. It undoubtedly explains the unrivaled success of this phylum among worms and helps to explain the extraordinary success of one of its relatives, the arthropods, which remained segmented even after the skeletal function of the coelom was lost.

Elastic skeletons do not change shape but simply bend when a muscle contracts. Muscle relaxation results either from a muscle contracting in the opposite direction to its antagonist or from the skeleton resuming its original position. The tentacles of many hydrozoan coelenterates, the mesoglea of jellyfish, the hinge of clamshells, and the notochord of chordates are examples. The high-pressured coelom contained in the rigid but flexible cuticle of nematodes also functions like an elastic skeleton.

Rigid, jointed skeletons achieve movement through a lever system. The elements of the skeleton are rigid segments attached together by flexible joints. Muscles span the joints and attach at each end to different elements. The more stable attachment site of a muscle is called the origin, the other the insertion. One muscle contracts and moves the skeletal element on which it is inserted, and an antagonistic muscle contracts and moves the skeletal element in the opposite direction. The biceps and triceps of the upper arm in humans are such a set of antagonistic muscles that bend and straighten, respectively, the lower arm. The control of movement can be quite precise with jointed skeletons. Muscles can bend or rotate skeletal elements whose length, shape, and number contribute to the resulting action. The dexterity of the hands is an example of the complexity of controlled movements made possible by a jointed skeleton.

Important to the speed and force of a movement are the length of the skeletal element and the size of the contracting muscle. Short limbs with thick muscles have more power than long limbs with slender muscles, but the latter have more speed. Limbs thus reveal a great deal about how an animal moves. Likewise, the relative massiveness of jaws reflects the toughness of the food eaten.

Two animal phyla, Chordata (vertebrates only) and Arthropoda, exploit jointed skeletons. Although the skeleton is internal in vertebrates and external in arthropods, the principles of movement are the same. A jointed skeleton is ideal for moving on land because adaptations for protection against dehydration (such as the cuticle) do not interfere with the action of the skeletal system. Indeed, the arthropod cuticle serves jointly a protective and a skeletal role. Moreover, the diverserange of precise movements made possible by this skeleton facilitates all sorts of locomotory patterns: swimming, digging, running, climbing, and flying. Jointed skeletons are also used directly for feeding (jaws). Arthropod jaws are derived from legs, while vertebrate jaws are derived from gill arches.

Translating movement into locomotion and feeding

Although all animals can move, not all locomote or displace the body over a distance. Locomotion serves the animal in finding food and mates and in escaping predators or unsuitable habitats. These functions of locomotion are typically correlated among different animals, so that those using the same mechanism of locomotion usually also feed, seek mates, and avoid danger in similar ways.

Some of the correlations between mode of locomotion and mode of feeding are described here, but space precludes discussion of the rich diversity found among animals past and present. The locomotory/feeding system of animals is the heart of their adaptation to their physical and biotic environments. Locomotory strategies for finding or gathering food include the following techniques.

Sitting still and waiting for food to arrive is particularly prevalent in aquatic habitats but is not rare on land. Sessile animals tend to develop strong defenses that are sometimes incompatible with effective locomotion. They rely on water or air currents or on the locomotion of their potential prey to bring food within reach. Because food may come from any direction, many sessile animals evolve radial symmetry. Settlement may be permanent or temporary, but in all cases one stage of the life cycle is capable of moving actively or passively from its place of origin. The choice of attachment site can also be active or passive; passive choice is often associated with an ability to grow in such a way as to maximize feeding efficiency. As with plants, passive settlers do well only with luck. The retention of locomotory capabilities requires energyand nutrients that can otherwise be diverted into growth or the production of offspring. Sessile feeders need to move if feeding and resting sites differ. Sessile animals include filter feeders, predators, and even photosynthesizers; the latter include corals that house symbiotic algae. Internal parasites are usually sessile because they live within their lifetime food supply. Mobile animals that pursue sedentary strategies for seeking prey include web-spinning spiders (a terrestrial mode of filter feeding) or deep-sea fishes with morphological adaptations that lure prey.

Burrowing animals typically eat the rich organic substrates they move through. Others burrow for protection and either temporarily emerge and gather organic sediments at the top of their burrows or pump water with potential food through the burrow. Instead of digging or finding burrows, some animals move into the centre of sponges, where they find protection and a renewing source of food.

Active movement in search of food requires energy, but this expenditure is more than made up for by an ability to seek out areas of concentrated food. This method of feeding applies to burrowing animals that eat the substrate through which they move, as well as to animals that move over solid surfaces, swim, or fly. Actively moving animals can feed on organisms that do not move, a rich variety coating virtually the entire solid surface of the Earth from the depths of the oceans to the peaks of many mountains. The main problem with this most productive avenue of food gathering is protection. Shells and poisons are the major types of defenses, although innovative detoxification metabolism and jaws of various kinds breach the defenses in part. This is an escalating battle, with the defenses, as well as the weapons to penetrate them, continually improving. Nudibranchs, shell-less marine snails, incoporate the defensive stinging cells of prey cnidarians into their own skin. Poisonous plants are eaten by specialized insects that avoid or detoxify the poison. In fresh water, for reasons not known, the arms race has not proceeded as far as in the sea.

Cooperation of individuals enables social animals to obtain food in novel ways. Uncannily like humans, some ants farm and herd other organisms for food. For example, some cultivate a fungus on leaves they cannot directly digest, while others herd aphids from which they milk nectar (actually the phloem sap of plants). Some ants even raid the nests of other species and make slaves of them. Another form of cooperation is the mutualism between species that trade advantage for advantage. Some fishes feed on parasites on the surfaces of other fishes, which benefits all but the parasites. In many animals, including termites and ruminants, microorganisms thrive in the gut and digest cellulose for them.

THE NERVOUS SYSTEM

Coherent movement results only when the muscles receive a sensible pattern of activating signals (for example, antagonists must not be activated to contract simultaneously). Animals use specialized cells called neurons to coordinate their muscular activity; nerves are bundles of neurons or parts thereof. Neurons communicate between cells by chemical messengers, but within a single cell (often extremely long) they can send high-speed signals through a wave of ionic polarization (analogous to an electric current) along their membranes, a property inherent in all cells but developed for speed in nerve cells by special modifications.

A system of communication requires three parts: a collector of outside information, an integrator to evaluate that information and decide upon its relevance, and a transmitter to convey the decision to the motor unit. In animals, sensory nerves and organs such as eyes collect the information; associative nerves usually concentrated into a brain integrate, evaluate, and decide its relevance; and effector or motor nerves convey decisions to the muscles or elsewhere. Although all three parts of the nervous system have kept pace with increases in the size and complexity of animals, the simplest systems found among animals (those of parazoans and coelenterates) are nevertheless capable of intricate feats of coordination. All ends of a coelenterate bipolar neuron can both receive and transmit an impulse, whereas the unipolar neurons of more derived animals receive only at one end (dendrite) and transmit at the other (axon). A neuron can have multiple dendrites and axons.

The earliest animals were probably radial in design, so that bipolar neurons arranged in a netlike pattern made sense. In such a design, a stimulus impinging at any point on the body can travel everywhere to alert a simple array of myofilaments to contract simultaneously. In the case of directed locomotion and relevant sensory input received at the head end of a bilateral animal, unidirectional transmission of nerve impulses to muscles becomes the only way to communicate effectively. The location of the brain in the head also reflects efficiency and the speed of receipt of information, because this position minimizes the distance between sensory and associative neurons as well as concentrates these two functions in a small, protected part of the body. In most animals nerve cells cannot be replaced if lost, although axons can be. Nerve cells tend to be concentrated centrally in ganglia or nerve cords, with long axons extending peripherally. Although certain animals may lose tails or limbs to predators or in accidents and then regenerate them, loss or damage to the central nervous system means death or paralysis.

The nervous system uses the transmission properties of neurons to communicate. Within a neuron, propagation of an impulse by an ion wave can be extremely rapid, but the wave can pass along the length of only one cell's membrane. To pass to the next cell at a synapse, where an axon meets a dendrite, a chemical transmitter is required. This molecule diffuses to the dendrites of a connecting neuron, where it initiates an ionic wave that propagates along the length of the cell's membrane. Although chemical transmission is considerably slower than the ionic wave, it is more flexible. For example, learning involves in part increasing the sensitivity of a particular nerve pathway to a stimulus. The sensitivity of a synapse can be altered by increasing the amount of transmitter released from the axon per impulse received, increasing the number of receptors in the dendrite, or changing the sensitivity of the receptors. Bridging the synapse directly by the formation of membrane-bound gap junctions, which connect adjacentcells, enables an impulse to pass unimpeded to a connecting cell.

The increase in speed of transmission provided by a gap junction, however, is offset by a loss in flexibility; gap junctions essentially create a single neuron from several. The same result can be achieved more effectively by lengthening the axons or dendrites, making some nerve cells metres in length. Situations arise where gap junctions become desirable, however. Gap junctions are found in vertebrate cardiac and smooth muscles, both of which transmit impulses along their cells to others. This ability makes these muscles somewhat independent of nervous-system control. A body can thus be kept partly functioning for some time without the activity of a brain.

Nerve impulses travel faster along axons of greater diameter or along those with good insulation against ion leakage (except at spaced nodes required for recharging). Vertebrates use their unique myelinated axons to increase the transmission rate of nerve impulses, whereas invertebrates are limited to using axons of greater diameter. As a result, vertebrates can concentrate more small neurons into a body of a particular size, with the potential for greater complexity of behaviour.

Memory is still a poorly understood aspect of the nervous system. As in learning, both short- and long-term memories seem to involve alterations in the ease with which subsequent impulses travel a particular pathway after it has been used. Transfer of memory through direct ingestion of the brain has not been confirmed experimentally. Although the underlying mechanisms are only dimly understood, it is known that there is a correlation between learning and memory capacity. The capacities for both increase with the number of associative neurons and the number of branches or interconnections formed. Since learning is a process of associating incoming cues with appropriate motor or internal response, greater memory capacity of a brain gives a more rapid learning process. Memory of inappropriate responses to an incoming set of cues can be used without motor repeat.

The degree to which the neurons of a brain develop interconnections is correlated with the complexity of its environs while growing. Consequently, a brain with fewer neurons but with more interconnections can be more "intelligent" than one with more neurons. Basic, repeated behaviours are inherited or learned by the development of fixed pathways by which an environmental signal reaches the motor nerves rapidly with little or no variation (reflex arcs). Nonreflex behaviour requires a decision to be made in the brain, with the resulting pathway to the motor nerves becoming more fixed (habitual) as one particular decision seems always to be correct. Reflexes are faster than decisions, but their relative adaptiveness depends on context. Animals vary in the degree to which they use reflexes or make decisions, patterns that are strongly correlated to brain size. Habitual actions are perhaps the most prevalent response, a compromise between the speed of a response and its appropriateness to context.

The senses

Appropriate behaviour relies on receiving adequate information from the environment to alert an animal to the presence of food, mates, or danger. Although sensory nerves carry this information to the brain, they do not always directly perceive the external world. Other modified cells intervene to convert light waves into vision, pressure waves in air or water into sound, chemicals into smell or taste, and simple contact into touch. Some animals have other senses, as for electric or magnetic fields.

In vision, for example, a photosensitive molecule changes shape and thereby sets off a chain of reactions that ultimately depolarize the dendrite of a sensory nerve. The associative neurons in the brain interpret the pattern of incoming impulses into a composite picture. What is "seen" may not entirely map what is really there: a great deal of filtering occurs, with editing by the brain to eliminate less important details so that only the most important are perceived. The accuracy of what is seen increases with brain size and the complexity of the visual gathering system, or eyes. Animal eyes range from being able to discern only the presence or absence of light to being able to see objects in vivid colour and great detail. Some animals see in ranges beyond unaided human vision. Pollinating insects in particular discern the colour of flowers differently than do humans; the ultraviolet reflection patterns of flowers do not always coincide with their coloured ones. Bees and birds perceive polarized light and can orient themselves by it. Some animals perceive long wavelengths, which are associated with heat (infrared), and can locate the presence of warm-blooded prey by such a mechanism.

Chemoreceptors are usually little-modified sensory neurons, except for the taste receptors of vertebrates, which are frequently replaced cells in synaptic contact with permanent sensory neurons. Chemoreception is based on the recognition of molecules at receptor sites, lipid-protein complexes that are liberally scattered on the dendrites of a sensory neuron. When the receptor recognizes one particular molecule by shape and sometimes chemical composition, it fires an impulse. The pattern of firings set off in the receptors of a certain molecule provides the information that the brain interprets as an odour or a taste. The details of how animals smell and taste are not as well understood as are the other senses. In many animals, chemoreceptors are not concentrated into obvious organs as they are in vertebrates, making even their location difficult to discern. Most animals possess some sort of chemoreception, and in many the sense is a major part of the animal's perception of its environment, far more so than it is for humans.

Sounds are waves of molecular disturbance that move through air, water, or solids, and their perception by animals simply uses sensitive mechanoreceptors. (Loud sounds can also be felt by the general touch receptors of the body and thereby influence its sense of well-being.) Sound receptors are sensitive hair cells or membranes that depolarize a sensory neuron when bent by the passage of a sound wave. Direct deformation of the dendritic membrane or release of transmitters by the hair cells fire the sensory neurons. Aside from a few insects, only vertebrates have organs with which to hear. Fishes and aquatic amphibians use a lateral-line system, and other vertebrates use ears; both organs use hair cells as phonoreceptors. Sound waves directly stimulate the hair cells of lateral-line systems, while sound waves only indirectly stimulate the hair cells of ears through an amplifying system of membranes and bones, which reaches a peak of complexity in mammals. Some animals (e.g., most bats and whales, and even whirligig beetles) use sound to "see" by echolocation. Sound is the preferred medium of communication between animals that hear. It can be used over longer distances than vision, and it can be used when vision is not possible. The signals decay more rapidly than do those of odours, and therefore the information can be more precise.

Mechanoreceptors also respond to touch, pressure, stretching, and gravity. They are located all over the body and enable an animal to monitor its state at any moment. Much of this monitoring is subconscious but necessary for normal functioning. Mechanoreceptors are often just sensory nerves, but other cells may be involved. Unlike other senses, that of touch is found in all animals, even sponges, where it reflects a general cellular trait of eukaryotes.

HORMONES

Hormones are the chemical integrators of a multicellular existence, coordinating activities from daily maintenance to reproduction and development. The neurotransmitters released by axons are one class of chemical communicators that act on an adjacent cell, usually a muscle cell or another neuron. Hormones are a mostly distinct class of chemical communicators secreted by nerves, ordinary tissue, or special glands; they act on cells far removed from the site of their release. They can be proteins, single polypeptides, amines, or steroids or other lipids. Hormones travel to their place of action via the circulatory system and then match their particular configuration with a specific receptor molecule attached to a cell membrane or, more usually, located within the cell.

The nervous system coordinates the more rapid activities of animal life, such as movement, while the hormones integrate everything else. Only the larger, more complex animals, such as vertebrates and some arthropods, have special endocrine glands to produce hormones; other animals use nerve cells or tissues such as the gonads. Endocrine glands are another example of a partitioning of functions into separate organs, a system that increases efficiency but that requires a relatively large size to maintain. Greater specialization is also associated with greater difficulties in regenerating lost parts or preventing breakdowns in functions.

Although the list of hormones found in the mammalian body may seem large, the numbers are surprisingly low for the variety of functions they influence. Which of the multiple functions any one hormone regulates depends on the specificity of the receptors on or within cells. Because all hormones bathe all cells as a result of their transport by the circulatory system, it is more efficient to have a general messenger transported to a cell, where it elicits only one of many possible outcomes. As in the nervous system, the specificity of response lies in the organ that responds and not with the messenger that merely commands action.

Chemicals that allow communication among individuals are called pheromones. Sexual attractants are the most common, but there are many other kinds.

Digestion

In contrast to plants, the essential nutrients that animals require to sustain life and to reproduce come packaged with their source of energy—the flesh or organic remains of other organisms. More complex animals tend to shorten and even eliminate many synthetic pathways, because most of the essential building blocks of their own complex molecules are present in their food. Reducing synthetic flexibility, however, inhibits a radical alteration in diet. The digestive and synthetic chemistry of animals strongly reflects their diets; some of this design may be altered with diet, and some may not. No matter how many leafy vegetables humans consume, for example, the cellulose remains undigested because appropriate microorganisms are not present in the digestive tract and they cannot be obtained at will. Consequently, essential nutrients are species-specific and tend to include only molecules adequately available in the usual diet.

The structure of a digestive system reflects its typical diet. Its purpose is to process food only to the point at which it can be transported to other cells for use as either fuel or structural material. In the simplest animals, such as sponges or some coelenterates, digestion is entirely intracellular, and some of the products of digestion are transported to nondigestive cells. As animals began to catch larger types of food, more of the digestive process had to be handled extracellularly. At the simplest level, seen in coelenterates or flatworms, large food items are held in an internal cavity (the gut) or even externally where certain cells release digestive enzymes. The food is broken down only to the stage at which it can be ingested by cells, which finish the process intracellularly. In more complex animals extracellular digestion accounts for virtually all breakdown of food before the products are transported to nondigestive cells.

Chemical digestion, whether intracellular or extracellular, is a relatively slow way to decompose a large item. Thus, animals begin to break it apart mechanically before exposing it to digestive enzymes. Teeth, the molluscan radula, and muscular gizzards are organs that speed up the digestive process by macerating food into finer particles.

Very early in their evolution animals acquired a one-way gut (gastrointestinal system), with the mouth typically armed with the macerating equipment and the terminal stretch sometimes specialized to retrieve excess water or other nutrients. Often a single passage through the digestive system leaves a great deal of useful material unclaimed. Because food moves along at a characteristic rate, which is sometimes influenced by how much is coming in, not all can be fully digested. Some animals regularly eat their feces to retrieve nutrients that may have escaped during first passage. If not recycled by their owners, feces are consumed by a diverse set of organisms.

A common specialization of the gut is the stomach or crop—a highly extensible part of the digestive tract that is used to hold a large amount of food and partially digest it before it enters the intestines, where most of the chemical breakdown and absorption of nutrients occurs. Most animals eat intermittently; the less often they eat, the larger the relative stomach size. Internalizing as much food as possible when it is available prevents potential food from being taken by a stronger competitor or enables a feeder to retreat to safety while digesting its meal. Ceca and second stomachs provide symbiotic microorganisms with a safe area within the gut to digest cellulose. Excess microorganisms mixed in with the partly digestible wastes contribute a steady protein-rich fare to the host in exchange for an optimal place to consume cellulose.

Stomachs predominate as a gut specialization because they allow animals to keep food from competitors or other dangers, but a few animals have developed ingenious methods of digesting their food before ingesting it. Humans are latecomers to this practice and have not yet carried it very far. Starfish exploit secondary radial symmetry and tube feet to open bivalved mollusks only enough to inject their stomachs, digest their meal within the protected shell, absorb the products, and leave the wastes behind. Spiders immobilize prey by silk wrappings and venoms, inject digestive enzymes, and drink the brew. Some primitive animals, like placozoans and certain flatworms, simply hunch over their prey as they digest it externally, a practice that leaves them vulnerable to other predators.

Animals use surfaces in many ways but no more strikingly than in the gut. Nutrients enter the body proper through the surface membrane of the gut; the larger the animal, the larger this surface area must be. The gut is probably the system that best reflects an animal's ecology. The simplest guts, found in animals from sponges to flatworms, simply branch like trees as the animal increases in size; the gut itself reaches all parts of the body to within the distance of a few cells and thus can serve for nutrient transport. As muscle masses become more prominent, the gut is squeezed into a more compact form. The gut compensates for this lack of space by internalizing its foldings. For example, the lining of the mammalian small intestine, the major site of digestion and absorption, is not only folded but each cell also has numerous outpocketings (microvilli), which increase the surface area 25-fold. Mammals and birds that primarily eat plants have longer intestines than those that favour meat. Warm-blooded animals, which maintain constant internal temperatures, require a great deal more energy than cold-blooded ones and thus tend to concentrate more surface area into a gut. Although they are not efficient energy users, it is to their advantage to obtain more usable energy even if efficiency is lost in the process.

WATER/VASCULAR SYSTEMS

Animals live in an aquatic environment even on land. Each cell is in contact with the ocean or its aqueous equivalent, which carries food and oxygen to the cells of the animal and carries its metabolic wastes away. The water/vascular systems found in animals vary from the nonexistent to the complex, with the complexity correlated with body size and level of activity. Smaller animals simply use the fluid-filled coelom for transport. Increasing size, however, places too many cells beyond diffusion distance from either the coelom or the outside. A muscular pump attached to muscular vessels has arisen in larger animals to move the interstitial fluid surrounding the cells. Most animals have open circulatory systems. Those few animals with closed circulatory systems have a continuous series of vessels to circulate fluid to the vicinity of all cells, whereas those with open systems have vessels only near the heart. (Actually, no system is entirely closed or open.) In open systems the interstitial fluid and the circulatory fluid are the same, but in closed systems the two fluids can differ considerably in composition.

Closed circulatory systems have several advantages that make them more appropriate than open systems for large, active animals: active animals, in fact, tend to possess closed systems even though their relatives may not. For example, cephalopods, alone among the mollusks, and nemerteans, the most active of acoelomates, have closed systems, as do all annelids and vertebrates. Decapod crustaceans, the largest living arthropods, have nearly closed systems. The most fully open systems have a heart with a few vessels leading from it, while fully closed systems both leak fluid (which is reclaimed by the open lymphatic system) and have open sections. For example, blood flow in the vertebrate liver is partly open.

In closed systems, blood flow can be both higher and directed more often to tissues that require a greater perfusion of blood. If blood is confined within discrete vessels, most of which are muscular, contractions can vary the flow rate according to need by altering the amount of constriction. Thus, the heart beats faster during exercise, when the muscles need more oxygen. Fear changes the distribution of blood flow to ready the muscles for possible imminent action. The more muscular arteries, which carry oxygenated blood to the tissues, can proliferate more finely in active tissues so that more cells are closer to the capillaries, where exchange takes place.

Another advantage of a closed system is the ability to carry a high density of oxygen-bearing cells. Such cells cannot flow smoothly through the sometimes tight interstitial spaces and thus are not much used by animals with open systems. A great deal more oxygen, however, can be carried if the oxygen carrier (such as hemoglobin) is packed into cells. The viscosity of the blood is a function of how many discrete particles are contained within it, and size is of little influence. If all the hemoglobin in the blood of humans were released by dissolving the cell membranes, it would be a thick gel unable to flow. Animals with open systems do aggregate their oxygen carriers into giant polypeptides, but single molecules have limits to their size. Myriapods and insects, highly active arthropods with open systems, circumvented this problem by evolving a tracheal system of respiration, as have some other groups: molecular oxygen is carried by branching tubes to within a few cell lengths of any cell.

A few types of cells protect organisms from a potentially hostile outside environment. Internal cells thus can eliminate any unnecessary ancestral life-support components as they specialize for various functions. This cooperation maintains an ideal internal environment for the members of the society of cells but only at the cost of active labour and expenditure of energy. In particular, the proper water/salt balance of the interstitial fluid is crucial to prevent the cell from shrinking or bloating.

Problems of water/salt balance are usually handled by the same system that eliminates the poisonous ammonia derived from metabolizing nitrogen-containing compounds, such as nucleotides or amino acids. Ammonia dissolves readily in water and thus is removed from an animal that needs to rid itself of excess water anyway. (In small animals the ammonia diffuses into the surrounding water.) With large size or a need for water conservation, animals excrete urea, a less toxic compound but one that also contains carbon and oxygen and thus potential energy. Urea also is highly soluble in water, but its low toxicity means that it can be concentrated before being excreted. Terrestrial animals with problems of water conservation either convert urea into uric acid, a solid compound that can be stored indefinitely in the body or voided with the feces, or develop efficient excretory organs (e.g., the mammalian kidney) that can concentrate the urea. Although water balance is usually handled by the kidney, salt balance is sometimes a specialized function of other organs. For example, because freshwater fish tend to lose a great deal of salt through their gills, they simply expend energy to concentrate salt against a gradient at this location.

Reproduction and life cycles

Primitive members of all major taxa of animals reproduced sexually, and virtually all animals still do at some time or another. In contrast to other activities, that of reproduction and life history may be most complex in the more simply structured animals. If little energy is put into complex maintenance systems, more is left for reproduction, the central focus of an animal's life. Thus, although locomotion constrains the reproductive strategy of an animal, the possibilities with any locomotory mode are diverse. For example, although sessile animals need not expend energy attracting a mate, they do face the problem of getting their gametes in contact with those of the opposite sex. Sometimes both sexes release gametes in immense swarms in which the probability of contact with the opposite sex is high. Often the female harbours large eggs, and the smaller, more mobile sperm are released to find them. In sponges, sperm simply enter with food. Hermaphroditism (the possession of both male and female capabilities) and parasitism by males are ways by which sessile, slow-moving, or sparsely distributed animals cope with finding mates. Barnacles, which are sessile crustaceans, elongate one limb to transfer sperm directly to another barnacle. (The hermaphroditism of barnacles lets any individual's neighbours be potential mates.) Some barnacles and other animals have small males that are parasitic on the females.

Mobile animals employ many kinds of devices for signaling their availability to the opposite sex. Pheromones, sound, and visual cues are used singly or in combination. Competition for mates may lead to elaborate courtship rituals, which enable a female to choose a suitable male; to size increases of males that fight for control of a harem; or even to size diminution and ultimately parasitism as males compete for a mate. In some species, sex changes with age, with males turning into females as they get larger. In a few animals, the sex depends on whether the individual settles on the substrate (becoming female) or on another individual (becoming a parasitic male).

Finding a mate is but one aspect of a reproductive strategy. The size of eggs is intimately related to the stage of development at which the young emerge to independent life, which in turn correlates with the habitat or mode of locomotion. For example, marine animals at one extreme produce vast numbers of tiny eggs, which hatch at an early developmental stage (e.g., the planula larva of coelenterates), or fewer, larger eggs, which hatch at a much later stage in the development toward adulthood. Smaller larvae spend more time feeding in the plankton before settling down to adult life, and during this time they are vulnerable to predation; however, they can disperse more widely, and their vast numbers give a positive chance that some will survive at each reproductive period. Terrestrial animals always produce relatively large, developmentally advanced young (spending the larval time in the egg), because the rigours of living on land demand immediately functional organ systems to sustain a free-living life.

Another problem faced by animals as well as plants is whether to breed only once during life, and thus to put all gathered energy into the effort, or to spend less energy during each reproductive period in order to grow and survive to reproduce for many years. A major factor affecting the evolution of a system of reproduction is whether the adult or the juvenile has the greater likelihood of survival. Some insects, such as mayflies, spend so little time as an adult (not much more than a day) that they have lost their feeding structures so as to allot more energy and space to reproduction. Breeding sooner means more descendants faster and more surely, so that mutations which are harmful late but helpful early are selected for. Therefore humans too senesce, unlike an amoeba.

Impler animals can pinch or bud off replicas of themselves, a mode of reproduction used by some animals that individually cannot get very large because of the simplicity of their organ systems. Such asexual reproduction is a form of growth but rarely of dispersal—the bud is usually sessile like the parent and thus remains adjacent to it. Mobility apparently requires such an integration of the nervous and muscular systems that it usually inhibits budding or fission.

Complex life cycles are an extreme variant on the usual life cycle of anmals. The juvenile or larval stage is simply more prolonged and complex; it is also structurally quite different from the sexually reproductive adult. Transformation to the adult may occur by asexual budding (e.g., coelenterates) or individual remodeling (e.g., insects or frogs). A complex life cycle enables an animal to feed in two different environments. It is not usually equally advantageous for the animal in both environments, so that one stage typically lasts longer than the other. For example, insects can become parasites without the usual problems of dispersal to a new host; the winged adult is admirably suited to find the correct host. Frogs can take advantage of ephemeralponds or ditches of water without competition from fish because in their terrestrial adult phase frogs can survive on land and thus locate new ponds when and where they become available. The cnidarian life cycle is also commonly one of alternation between a mobile and sessile form. Some animals alternate between reproduction from unfertilized eggs (all females) and sexual reproduction. The all-female generations can reproduce faster to take advantage of seasonally excessive resources (e.g., aphids or many freshwater crustaceans).

Systemic estrogen has many effects on organ function and blood flow that vary depending on the individual vascular bed being studied. Female reproductive tissues like the ovary, vagina, and uterus are exquisitely sensitive to the vasodilatory effects of estrogen.[1–3] Uterine vascular bed responses are partially reduced by giving combination treatment with progesterone. In numerous species including sheep, the systemic estrogen to progesterone ratio is elevated during the follicular phase when uterine blood flow is greatest, while during the luteal phase, the estrogen to progesterone ratio is depressed and uterine blood flow is at its nadir.[7] Pregnancy is characterized by high endogenous estrogen levels in a progesterone-dominated state with substantial rises in uterine blood flow.

The relationship between systemic estrogen replacement and bladder volumetric perfusion has been studied These studies show an increase in bladder perfusion with estrogen replacement therapy (ERT) and a fall in blood flow with oopherectomy. Bladder blood flow (BBF) has been estimated using a variety of different techniques, including Doppler ultrasound flow velocity, O_2 consumption and microspheres (florescent and radioactive). There have also been studies confirming the presence of estrogen receptors (ERs) in the bladder and the effects of ERT on receptor levels during menopause.

Numerous functional studies of the bladder have evaluated estrogen replacement and contractile function. They noted an increase in contractile function of bladder smooth muscle with estrogen treatment. In vivo tissue oxygenation of the bladder is also increased with ERT as well as increases in angiogenesis, changes in bladder structure, and detrusor muscle density in rabbits with ERT.

To our knowledge, the effect of endogenous estrogen on BBF and bladder ERs during the ovine ovarian cycle has not been studied. There have been several studies examining urethral blood flow during different phases of the menstrual cycle using Doppler ultrasound. No statistically significant differences were observed. The changes in BBF during pregnancy are not known. We believe that alterations in endogenous steroids of ovarian origin may play a role in the maintenance and/or health of the bladder and the tissues of the pelvis. To our knowledge, this is the first study directly evaluating BBF in sheep in different endogenous and exogenous hormonal states using blood flow measurements.

The primary objective of this study was to quantify the effects of endogenous and exogenous estrogen on bladder vascular resistance and BBF using a physiologically well-defined animal model for examining endocrine states of high and low estrogen. The secondary objective of this study was to determine the levels of ERá, and ERâ in the bladder during these different hormonal states. We hypothesized that during physiological states of high endogenous estrogen (follicular phase and pregnancy) or exogenous ERT, BBF will increase and with oophorectomy as a surgical model of menopause, (i.e. with low estrogen), BBF will be reduced. Moreover, we hypothesize that ER levels will directly correlate with these changes in BBF.s

Prior to surgery, the animals were fasted and water removed for 12 to 24 hours. Anesthesia was induced with an intramuscular (IM) injection of ketamine (15 mg/kg; Midwest Vet Supply, Sun Prairie, WI) and atropine (12 mg/kg, Hospira Inc., Lake Forest, IL). Prophylactic IM antibiotics (400 000 U penicillin; Sandoz, Inc., Princeton, NJ; 4 mg/kg Gentamicin, Abraxis, Inc., Schaumburg, IL) were also given. A jugular vein catheter was inserted and advanced 20 to 30 cm to administer a continuous IV drip of 10%ketamine in 0.9% saline and 5% dextrose. Supplemental Nembutal (pentobarbital sodium 50 mg/ml; Sigma Chemical Co., St. Louis, MO) was given as needed throughout the procedure to insure an adequate plain of anesthesia was maintained. Surgeries were performed on 47 mixed western breed sheep (62 ± 3.2 kg; 6 pregnant 120–130 days' gestation [term = 147 days; n = 6], 41 nonpregnant), which consisted of implantation of arterial and venous catheters inserted in both superficial saphenous vessels and advanced through the femoral circulation into the descending aorta and abdominal inferior vena cava, respectively. In addition, a left ventricle catheter was implanted via the left carotid artery. These catheters were exteriorized and maintained under heparin lock (100 IU/mL). Thirty of the nonpregnant animals had bilateral oophorectomy by midventral laparotomy. Postoperative 75 mg IM Flunixin Meglumine (Banamine, Schering-Plough, Palo Alto, CA) analgesia was administered immediately postsurgery and again at 24 hours. Animals were allowed access to food and water ad libitum throughout the remainder of the study. All animals were given IM penicillin (400 000 U) and Gentamicin (4 mg/kg) for 2 days and then every 3 days thereafter. The animals were returned to the laboratory on postoperative day 4 to 5 following surgery.

BLADDER BLOOD FLOW MEASUREMENTS

Experiments to measure BBF were conducted at least 5 to 7 days postsurgery. In these studies, mean arterial pressure and heart rate were monitored continuously for 60 minutes, arterial pH, PO_2, PCO_2, hemoglobin, hematocrit, lactate, and glucose levels were measured intermittently, and all were in normal physiological ranges. An injection of a randomly selected 15 um radioactive (^{153}Gd, ^{51}Cr, ^{113}Sn, ^{85}Sr, ^{95}Nb, and ^{46}Sc) microspheres (3–5 million spheres/injection) was given via the left ventricular catheter, while a blood reference sample was withdrawn (4.12 mL/min) from the descending aorta (via both saphenous artery catheters)· To establish time courses of changes in BBF relative to expected changes in estrogen and/or progesterone levels, hemodynamic monitoring and Microsphere injections were repeated.

Experimental Protocol for Intact Nonpregnant Cycling and Pregnant Animals

Estrous behavior in the nonpregnant intact sheep (n = 11) was determined using a vasectomized ram, and surgery performed on day 3 to 4 post estrus.[7] On day 9 postestrus, the sheep was given 2 IM injections 4 hours apart of either lutalyse (follicular group—1 mg $PGF_{2á}$, Upjohn Pharmaceuticals, Kalamazoo, MI; n = 7) or saline (luteal group—1 mL; n = 4). Microsphere injections and reference samples were evaluated immediately prior to lutalyse or saline injection and at 6, 24, 44, 46, and 48 hours. There were no time-dependent changes in BBF in the saline controls; therefore, luteal phase BBF calculations were averaged within an animal and were based on all time points for the saline-injected sheep (day 5, 9–11) and pre-lutalyse injection (day 9). Follicular phase BBF from the lutalyse-injected animals was the average from 44 to 48 hours (day 11), which is during the peak estrous cycle periovulatory period when both the estrogen to progesterone ratio and the uterine blood flow are maximal.Pregnant sheep were studied at 120 to 130 days' gestation (term = 147 days; n = 6) and were given an injection of radioactive microspheres as described above after a control period of 60 to 120 minutes of established normal resting blood pressure, heart rate, and arterial blood gases.

Experimental Protocol for Oophorectomized Vehicle Versus ERT Animals

Oophorectomized (Ovx) sheep were monitored for 45 to 60 minutes to confirm resting blood pressure, heart rate, and arterial blood gases at which time the first micro-sphere injection was performed in 30 non-pregnant Ovx sheep. A subset of these sheep were given IV injections of either vehicle (10% ETOH in 3 mL saline; n = 7) or E_2â (5 ìg/kg bolus over 1–2 minutes; n = 9) followed immediately by 5-mL saline flush. Two hours after the E_2â injection another microsphere injection was performed for blood flow measurements (ERT acute). Vehicle (95% ETOH) or E_2â (6 mg/kg over 24 hours) infusions using Harvard syringes infusion pump (0.009 mL/min) were continued IV for 10 days in the same sheep, with additional microsphere injections and reference samples obtained at 3, 6, 8, and 10 days (ie, chronic, vehicle, or ERT, respectively).

Tissue Collection and Processing

After the last microsphere injection was performed, sheep were euthanized with an overdose of pentobarbital sodium (50–70 mg/kg). Tissues were weighed and placed into vials, and the level of radioactivity for each isotope per microsphere was determined as previously described.[1–3,32] Bladder blood flow was calculated using following standard equation (withdrawal rate/# of microspheres in reference blood) (# of microspheres in the tissue sample). Resistance was calculated as mean arterial blood pressure/regional blood flow. All bladder tissues had >400 microspheres/bladder.

Western Blotting for ERs Concentration

Bladder tissue from the dome of the bladder of 40 additional sheep was collected and snap frozen in liquid nitrogen at the time of sacrifice. All samples were stored at ("4 C) and subsequently were processed together for protein assay, Western blotting, probing, and analysis. The bladder tissues so obtained were matched to the following physiologic groups evaluated above for BBF (10 luteal, 8 follicular, 8 pregnant, 10 Ovx/Vehicle, 4 Ovx/Chronic Estrogen). Estrogen receptor á and ERâ levels were evaluated with anti-human ERá and ERâ antibodies (Neomarker, Fremont, CA and Santa Cruz, CA respectively).[33] Blots were designed with a pooled standard sample used to normalize differences in exposure or antibody binding. After enhanced chemolumenience probing, blots were scanned and analyzed with densitometry.

STATISTICAL ANALYSIS

Data were analyzed using a single tailed Student's t test. These data were analyzed as absolute flow (mL/min/g) and vascular resistance (mm Hg/mL/min/g), and values are reported as means ± SE.

We observed that the dome of the bladder had primarily ERá with negligible levels of ERâ. These findings are consistent with other studies that demonstrate ERâ levels in the trigone of the bladder and urethra but not in the dome.[19] We selected this tissue site because overactive bladder symptoms in women are related to abnormalities in function of the dome of the bladder.[34–38]

We observed that BBF and ERá levels were not significantly changed with fluctuations in endogenous estrogen levels during the ovarian cycle (Figure 1). This is also supported by studies measuring bladder blood velocity using Doppler ultrasound during the menstrual cycle in women,[13,30] which suggested little if any variation in volumetric BBF. This is substantially different when compared to uterine tissue and other reproductive tissues, which exhibit substantially increased blood flow during high endogenous estrogen states such as the follicular phase of the ovarian cycle.[2,4,7,8,32] Indeed uterine tissue blood flows were elevated greatly during the periovulatory period in uteri collected from the same animals used for the current study (data no shown).

We observed that BBF levels fall and unexpectedly ERá levels rise with oopherectomy .The fall of BBF and rise in ERá levels with oopherectomy suggest that decreased circulating estrogen levels and/or decreased BBF lead to a rise in ERs. When estrogen was replaced in a chronic manner, we observed a rise in BBF that approached levels seen in cycling follicular and luteal sheep. Our observations are consistent with data obtained from humans and other animal models on ERT.The BBF responses to systemic estrogen in this study are also consistent with functional studies in humans and animal models, which show improved bladder function and in particular contractile activity.In the present study, we noted a significant rise in ERá levels with Ovx that was unexpectedly unchanged with chronic ERT. The reason for this is unclear and needs further investigation. This implies that unlike reproductive tissues, factors other than tissue estrogen levels appear to modulate ER levels in the bladder.

We recognize and wish to point out a limitation of the current study compared to the human, in that all surgically menopausal sheep were not aged and were oopherectomized for only 7 to 15 days to achieve this state. Most symptomatic postmenopausal women have intact ovaries. It is possible that compounds produced by the ovary other than estrogen participate in modulating BBF. It is also possible that other diseases of aging may affect BBF in postmenopausal women or that progesterone may possibly be an important factor. In this regard, preliminary data evaluating the effects of exogenous progesterone in Ovx sheep suggest that like the uterine circulation the acute (2 hours) estrogen BBF response is somewhat attenuated by progesterone. However, in Ovx sheep no overt BBF differences were noted in the chronic progesterone treatments compared to either chronic estrogen or vehicle alone (data not shown).

Several laboratories have studied the physiologic significance of blood flow changes as a result of estrogen. It is noteworthy that relative changes in blood flow in response to estrogen have been shown to be much greater in the reproductive vasculature than the systemic vasculature. The urogenital tissues and the tissues of the urogenital sinus have been show to be estrogen responsive whereas there are multiple tissues in which estrogen does not have any effect what so ever.[1] To this end, comparing blood flow during different hormonal states in any of the nonreproductive tissues is important. Other animal models have measured BBF; in Parekh et al[11] blood flow was measured in female rabbits where comparisons were made between intact and Ovx animals. Indeed their comparisons like these described herein also found significant reductions in blood flow with Ovx. They noted a 50% Ovx-induced reduction in blood flow to both the detrusor muscle and the urothelium, translating to a change in flow of 0.025 mL/min/g for detrusor muscle and a change of 0.17 mL/min/g for urothelium. In our study, full-thickness bladder samples were taken that included urothelium and detrusor muscle; we noted a 62% change in flow between intact (luteal) and Ovx ewes that translates to a change in flow of 0.113 mL/min/g. No literature comparisons are available for Ovx vehicle-treated versus Ovx ERT-treated (acute or chronic) or hormonally intact versus pregnant animals.

Although functional studies (eg, bladder contractility, capacity, etc) have not been performed as part of this investigation, there are functional studies in the literature in the premenopause, post menopause,[11] and postmenopause with ERT groupsAll these studies confirm decreased function with lower estrogen physiological states and a return of function with estrogen replacement. There are many bladder function surveys in women during these different hormonal states that again confirm the clinical significance of these findings.

During pregnancy, we unexpectedly observed a significant decrease in BBF and a rise in ERá levels. It is possible that this decrease in blood flow is due to shunting of blood from the bladder to the adjacent reproductive tissues. This would likely occur because the reproductive organs are much more sensitive to the vasodilatory effects of estrogen and have a relatively much greater fall in vascular resistance than the bladder. It is also possible that other regulatory factors may play a role in modulating the responsiveness of this organ during gestation. The ratio of estrogen to progesterone and the timing of exposure to these steroids is dramatically different during pregnancy than the ovarian cycle. Pregnancy is a chronic progesterone dominant state, but how these pregnancy levels or exogenous progesterone treatments versus luteal phase exposures affect BBF is unclear. It is of interest that functional studies in animals show that pregnancy improved bladder capacity and compliance.This is thought to be due to high circulating progesterone levels that lead to structural changes in the detrusor muscle. Functional studies in humans are less clear but do not show a decrease in bladder capacity

We also considered that an increase in urine production during pregnancy due to increased glomerular filtrations rates may be the cause of decreased blood flow as well as frequency and urge symptoms. Although this is briefly commented on for completeness sake, we do not feel very strongly that this is the exact mechanism for the fall in BBF in our model. In animals, this has been shown to be true if there is evidence of poor bladder compliance. If the bladder has normal compliance, then studies have shown no decrease in blood flow during filling. Because we expect that pregnancy improves bladder compliance, it is unlikely that increased urine production is the true mechanism for these symptoms. Many clinical studies reveal that regardless of any changes in capacity and compliance, pregnant women uniformly complain of a marked increase in overactive bladder symptoms as early as the first trimester. Our findings of a fall in BBF during pregnancy may partly help explain this.

In summary, the bladder appears to respond to endogenous ovarian estrogen levels for tropic support of basal function rather than the cyclical changes seen in uterus and other reproductive organs. It is likely that during gestation the observed decreases in BBF are associated with the substantial alteration in pelvic organ hemodynamic redistribution and may contribute significantly to the clearly elevated overactive bladder symptoms reported throughout gestation.

We have previously observed that exogenous oestradiol can delay the intercycle increase in plasma follicle stimulating hormone (FSH). The increase in plasma FSH that follows discontinuation of exogenous oestradiol peaks after 3 days. We have now studied the possibility of using exogenous oestradiol to synchronize the increase in endogenous FSH with the onset of human menopausal gonadotrophin (HMG) treatment in controlled ovarian hyperstimulation (COH). A total of 30 women aged 35.1+/-6.3 years (mean+/-SD) undergoing ovarian stimulation received 2 mg of oestradiol valerate twice daily starting on day 25 of the previous menstrual cycle until the first Tuesday following menses. Ovarian stimulation was initiated 3 days later. On the last day of oestradiol treatment, plasma oestradiol, FSH and luteinizing hormone (LH) (mean+/-SEM) were 566+/-53 (pmol/l), 3.8+/-0.4 (IU/l) and 5.5+/-0.8 (IU/l) respectively. After 3 days, the FSH and LH (mean+/-SEM) had increased to 6.7+/-0.7 and 6.9+/-0.7 (IU/l) respectively while oestradiol decreased to 251+/-29 (pmol/l). The mean number (+/-SEM) of HMG ampoules used was 25.1+/-2.7 and treatment lasted 11.3+/-0.9 days. Five women became pregnant for a pregnancy rate (ongoing) of 19 (15)%. If all women aged >40 years (six women who did not become pregnant) were excluded from analysis the pregnancy rate (ongoing) was 24 (19%). These results indicate that exogenous oestradiol can safely be used for the synchronization of endogenous and exogenous FSH stimuli in COH. This approach provides the practical advantage of permitting an advanced timing of the onset of COH treatments when gonadotrophin-releasing hormone (GnRH) agonists are not used, which improves treatment convenience for patients and team members alike. Further development of this model may enable control of the onset of natural cycles which may find practical applications for timing assisted reproductive techniques (intrauterine insemination or in-vitro fertilization) in the natural cycle.

CONTROL OF OVULATION IN FARM ANIMALS

Examination of hormonal changes occurring in farm species at the onset of puberty, during the follicular phase of the oestrous cycle, and at those times when ovarian activity is re-established after periods of seasonal or lactational anoestrus, provides circumstantial evidence that the final phases of follicular development are dependent on a pattern of tonic (episodic) LH secretion. A suppression of episodic LH secretion is associated with periods of anovulation. Stimulation of tonic LH secretion by repeated injections of small doses of synthetic Gn-RH or purified LH restores normal reproductive function in all but deeply anoestrous animals. Continuous infusion of Gn-RH is as effective as repeated injections. It is suggested that an additional inadequacy, possibly endocrine, contributes to the anovulatory state in deep anoestrus.

Physiological and Reproductional Aspects of Animal Production

In mature follicles just prior to ovulation, ova are usually seen surrounded by a halo of granulose cells (cumulus) that are continuous with granulose cells lining the fluid-filled antrum. The large, thin-walled follicles bulge from the ovarian surface . The primary oocyte, which remains in an arrested stage of meiosis during follicular development, undergoes the first meiotic division to produce a secondary oocyte and first polar body just before ovulation in most species. The first polar body is extruded from the ovary with the secondary oocyte. In the mare, the first meiotic division occurs just after ovulation.

This short–term change in LH release is the LH surge. As discussed earlier, the LH surge depends on changes in the hypothalamic-hypophyseal axis and increase in adenohypophysis content of LH induced by the rapid rise in production of estrogens by large, mature follicles. The extremely high levels of LH promote development of the primary oocyte and its progress through the first meiotic division. This prepares oocyte for ovulation.

Granulosa cells also respond to the LH surge by transforming from estrogen-producing cells to progesterone-producing cells. This is part of luteinization, the transformation of granulosea cells to luteal cells (cells of a corpus luteum). This process begins prior to ovulation, so estrogen levels are decreasing and progesterone levels are increasing at ovulation. Under the influence of the LH surge, granulosa cells also acquire the ability to synthesize prostaglandins, thromboxanes, and leukotrienes. These agents induce a local response similar to inflammation that will weaken the wall of the follicle and promote rupture.

Spontaneous and reflex ovulation

The LH surge and ovulation occur in most domestic species (mare, cow, ewe, and sow) independent of copulation, ant these species are spontaneous ovulators. In these species, the preovulatory increase in estrogens from developing follicles is the primary event that brings about ovulation. The female animals of some species (rabbit, ferret, mink, camel, llama, and alpaca) usually require copulation for ovulation. These are induced ovulators. In these species, the final preovulatory surge of GnRH, and subsequent LH surge, is apparently dependent on a neural reflex elicited by vaginal stimulation. Induced ovulators have characteristic estrous cycles and follicular development, but mature follicles regress if copulation does not occur.

BREEDING SEASON

The term breeding season refers to the period of time during the year when females of a species come into estrus. Most species of wild animals have a definite breeding season which is initiated at a time when environmental conditions are conducive to the survival, growth, and development of the young one after birth. For example, the gestation period in sheep and goats is 144 - 152 days, and the breeding season in these species starts in fall so that the young are born in spring. In cattle, where pregnancy on average lasts for nine months, the female breeding season occurs in the early summer so that the calves are born in the spring when the temperature is optimum for the growth of the newborn. Green fodder and forages are also available, which support sufficient milk production by the dams.

Domestication of farm animals has resulted in some variation of breeding season. Due to the intervention of modern production techniques, farm animals are now capable of breeding throughout the year. Among the farm animals, cows and buffaloes are polyestrous, showing diestrous cycles throughout the year. Mares, ewes, and does are seasonally poly estrous, showing diestrous cycles during a definite breeding season. The breeding season of ewes and does is confined largely to fall and winter. In Indian sub continent, buffaloes breed from October to December, when day length is decreasing. The stimulus for the initiation of reproduction during certain seasons results from the action of light, via the eye and optic nerve, on the pituitary gland. This stimulus causes the pituitary to release gonadotropic hormones, which activate the functioning of the gonads. One of the limiting factors influencing the productivity of the buffalo is the seasonality of breeding, which may be due to either anestrus or silent estrus during the hot, dry months of summer.

Estrus Cycle

Estrus occurs in non-pregnant females in characteristic rhythmic cycles. The interval between two consecutive estrus periods is known as the estrus cycle. The estrus cycle in farm animals is characterized by cyclic changes in the morphology of the reproductive organs as well as in the behavior of the animals.

The estrus cycle is divided into four different phases; these are known as proestrus, estrus, metestrus and diestrus.

Proestrus is the phase of estrus in which the reproductive system is beginning preparation for the release of the mature ovum from the ovary. Estrus is defined as that phase of the cycle when females are receptive to coitus. Metestrus is the short transitional stage following ovulation in which the effects of estrogen are declining. The recently ruptured follicle is reorganized, and the secretion of progesterone is rising. During this phase, the uterus gets prepared for the implantation of the embryo. If pregnancy does not occur, the reproductive tract goes into a period of rest called diestrus. In cows and buffaloes the estrus cycle averages 21 days in length. Sheep and goats usually have a shorter estrus cycle ranging from 16 - 21 days.

Signs of estrus are somewhat similar in females of different species. In cows and buffaloes, estrus is characterized by the manifestation of heat. The cows may bawl frequently, become very active and restless, or mount other females and remain mounted. The vulva swells and the vestibule may be come deep-red. The swelling subsides with the end of estrus. There is marked discharge of clear, viscid mucus secreted by the cervix from the vulva. The excessive secretion at estrus acts as a lubricant during copulation.

Ovulation: Ovulation may be defined as the discharge of the ova from the Graafian follicle. Ova develop from germ cells in the ovaries. Numerous germ cells are present in the ovaries at birth, but few of the follicles surrounding them mature to rupture and shed ova. Follicles grow under the influence of the follicle stimulating hormone (FSH). The layers of follicular tissue become thinner with growth, the follicle rupturees and ova are released. This process is stimulated by the luteinizing hormone (LH). The released ova are swept by muscular and ciliary movements into the oviduct towards the uterus. If mating or articficial insermination takes place, spermatozoa and ova meet in the oviduct.

Esturs and ovulation are more or less synchronized in females to increase the chances of fertilization, the union of an ovum with a spermatozon. Synchronization of ovulation and insemination is essential because the life of the ovum once it is shed and that of the spermatozoa in the female reproductive tract is limited to a few hours.

Gestation: Gestation is the process that begins with the fertilization of the ovum by the sperm and terminates with the birth of a young one. The period of gestation varies considerably among species of farm animals. The average gestation period of cattle is 282 days, buffaloes 310 days, and sheep and goat 151 days. Generally, male calves are carried one day longer than female calves.

Parturition: Parturition is the physiological mechanism that enables the uterus to expel fetus after a certain period of development and nourishment. The process is called "foaling" in mares,"calving" in cows and buffaloes, "lambing" in ewes, and "kidding" in goats.

Signs of Parturition. The dam shows a number of signs of approaching parturition, which are usually divided in to following four stages.

(a) Preliminary stage: This stage may last for hours or even days. The ligaments in the pelvic region and around the tail head begin to relax and sink a few days before calving. The dam exhibits difficulty in movement. The external genitalia are swollen, enlarged, and flabby, and strings of mucous are seen. The mammary glands are enlarged and become tender. A clear, waxy fluid oozes out of the teats when pressure is applied.

(b) Dilation of the cervix: This stage merges with preliminary stage. Uneasiness increases and the animal may lie down and get up frequently. Frequency of micturition (urination) increases. At this time the cervix is fully dialted, and the next stage follows without any appreciable break in the sequence of events.

(c) Expulsion of the fetus: This staage begins with the fetus entering the dilaated cervix and the birth canal. During this time, uterine contractions occur at about two-minute intervals and the amnion is not ruptureed. There is a rush of fluid from the uterus when the water bag rupturees, and the animal appears relieved. Normally, the forefeet, with the muzzle lying behind and over them, appear at the vulva. The young one is expelled from the fetal membranes by a forceful and painful effort by the mother.

(d) Expulsion of membranes: The expulsion of the fetal membranes is commonly known as the delivery of the afterbirth. COntinued contractions of the uterus disjoin the fetal cotyledons, and the placenta is expelled within half an hour to eight hours after partuirition.

Roles of exogenous and endogenous FGF-2 in animal models of depression

PURPOSE:

Several members of the fibroblast growth factor (FGF) family have been shown to be dysregulated in individuals with major depression, and treatment with antidepressants has been reported to increase FGF-2 mRNA levels in the forebrain.

METHODS:

We have used the tail suspension test (TST), and olfactory bulbectomy (OBX), and FGF-2 deficient mice to investigate putative roles of FGF-2 as an antidepressant and mediator of antidepressive drug actions.

RESULTS:

FGF-2 applied intraventricularly generated antidepressant-like effects in the TST. FGF-2, similar to the antidepressant amitriptyline, attenuated neuron demise in the piriform cortex and posterolateral cortical nucleus of the amygdala following OBX. Moreover, OBX induced reduction in hippocampal neurogenesis could be ameliorated by subsequent treatment with either amitriptyline or FGF-2. Furthermore, FGF-2 was effective in reversing depressive-like behavior induced by OBX, monitored in the locomotor activity and the passive avoidance test. In bulbectomized FGF-2 deficient mice, treatment with amitriptyline protected neurons, but failed to reverse behavioral alterations.

CONCLUSIONS:

Together, these results suggest that FGF-2 constitutes both a potential target for antidepressive treatments and an important growth factor in the cytokine network underlying the actions of antidepressive drugs. The results further suggest a requirement of endogenous FGF-2 for mediating behavioral, but not neuroprotective actions of amitriptyline.

Anyone who has lived for any time in the world recognizes that the world changes in cyclical patterns: day and night, the seasons, and tides. Many animal species have a sort of internal clock, called a biological clock, which predicts cyclical environmental change and prepares the animal to deal with it. Biological clocks can be set by exogenous (external) stimuli, or by endogenous (internal) rhythms. Exogenous stimuli are called Zeitgebers, which means "time giver" in German, and include light, temperature, and length of day. Endogenous clocks are responses to an internal rhythm that is pre- programmed to correspond with the environmental temporal pattern.

CIRCADIAN RHYTHMS

Circadian (literally "about a day") rhythms are endogenous animal clocks that operate on a daily time schedule. Most circadian rhythms are actually slightly shorter than 24 hours, some are longer, and a few are exactly 24 hours. Jet lag provides a common example of circadian rhythm. When a person experiences jet lag, their biological clock is alerting them that it has reached the time to sleep during their twenty-four hour cycle, but because they have traveled so far the internal clock mechanism does not coincide with the actual time of day.

In general, circadian rhythms are endogenous, acting independently of external stimuli. Most introductory tests (such as AP Biology) will probably treat circadian rhythms as such. However, some daily cycles are maintained by a combination of an endogenous clock and external stimuli such as sunlight. In Drosophila (fruit flies) rhythms for determining emergence from the pupal stage, which must take place at a specific time of day, are determined by two gene products, Per (period) and Tim (timeless). Per is involved in a self-regulatory feedback loop which causes cyclic changes in its concentration over the course of the day. Tim is also self-regulatory, but it is also light sensitive, so changes in the normal light-dark cycle will be reflected in Tim concentrations. These two proteins constitute a daily clock that is both endogenously and exogenously regulated.

The dependence of some internal clocks on the presence or absence of light presents an interesting problem: how do cells deep within the body know if it is light or dark out? In vertebrates, this information is routed through an organ called the pineal gland. In some fish, birds, and reptiles this organ has photoreceptors and can sense light directly. In mammals the pineal gland has no light sensing ability of its own, but rather receives light information from the eye through the suprachiasmatic nucleus (SCN), which connects to the pineal gland through the sympathetic nervous system. The pineal gland secretes proteins such as melatonin, which send the light signal to the rest of the body.

Circannual and Lunar Rhythms

While behaviors such as pupal emergence, sleep, and activity must be regulated on a daily basis, other behaviors need to be regulated over longer periods of time. For example, mating in marine organisms is often linked to stages of the tides. In the paolo worm, mating occurs in the neap tides during the last quarter moon of October in November. These worms are very vulnerable during mating, so by synchronizing their mating behavior to an exogenous factor such as tides, they can outnumber predators and increase the probability that some mating will be successful. This circannual mating rhythm provides the paolo worm with increased chances of survival: it makes both individuals and the species more fit. In other words, the rhythm is selected for by natural selection, and thereby evolves into a common species trait.

Divergent Effects of Endogenous and Exogenous Glucocorticoid Induced Leucine Zipper in Animal Models of Inflammation and Arthritis

Profound antiinflammatory actions of glucocorticoids are the basis of their widespread use in rheumatoid arthritis (RA), despite adverse effects reflecting their metabolic actions .Greater understanding of mechanisms may enable the development of therapies that mimick the antiinflammatory, but not the metabolic, effects of glucocorticoids. Glucocorticoid actions are integrated at the molecular level by the effects of the glucocorticoid–glucocorticoid receptor (GR) complex on gene transcription . The glucocorticoid–GR complex binds as a dimer to glucocorticoid response elements in target gene DNA, or inhibits ("transrepresses") gene expression by tethering as a monomer to other transcription factors such as NF êB and activator protein 1 (AP 1) (5). Among the genes most sensitive to transcriptional activation by glucocorticoids is glucocorticoid induced leucine zipper (GILZ) (also called TSC22D3) . GILZ inhibits NF êB through a physical interaction with the NF êB p65 subunit which impedes nuclear translocation (7). GILZ also binds to the AP 1 components c Jun and c Fos and inhibits Raf 1 and Ras . GILZ silencing increased the severity of murine collagen induced arthritis (CIA), and GILZ overexpression inhibited chemokine and cytokine expression in human RA synovial fibroblasts (RASFs) . These observations have led to the concept that GILZ is a critical mediator of the antiinflammatory effects of glucocorticoids.

The majority of the effects of GILZ have been reported in forced overexpression studies. Although male infertility was recently reported in a GILZ knockout (KO) mouse strain , the physiologic effects of GILZ on the immune system, and whether GILZ is required for the therapeutic effects of glucocorticoids, remain unknown. To investigate these issues and to probe the potential for GILZ based RA therapies, we generated GILZ″/″ mice and the means to locally overexpress GILZ in vivo. We report that although GILZ deficiency resulted in increased T cell activation, there was no effect on effector pathways of inflammation or on the expression of RA in 2 distinct models. Moreover, and unexpectedly, glucocorticoid effects on inflammation were retained in the absence of GILZ. In contrast, therapeutic overexpression of GILZ mimicked the inhibitory effects of glucocorticoids in a model of RA. These findings revise our understanding of the role of GILZ in immunity and inflammation and in the actions of glucocorticoids, yet they support the possibility of GILZ manipulation as a therapy for RA.

MATERIALS AND METHODS

GILZ deficient mice were generated (in conjunction with Ozgene), by disruption of the Gilzgene in C57BL/6 mouse embryonic stem cells via homologous recombination using the Cre/loxP system.The targeting construct of the Gilz gene was built by using the in house targeting vector backbone (FLSniper; Ozgene) with homologous sequences spanning <"6 kb upstream and <"6 kb downstream of the Gilz gene. The floxed genomic area was <"2.2 kb, with two loxP sites flanking the last exon of the Gilz gene. The neomycin cassette was removed by crossing with Flp deletor C57BL/6 mice. The resulting heterozygous floxed animals (GilzfloxedÄNeo/wt) were crossed with a general Cre C57BL/6 deletor, and the resulting Gilzflox/wt, Cre+ mice were backcrossed to wild type (WT) C57BL/6 mice to remove the Cre gene.

Induction of delayed type hypersensitivity (DTH), antigen induced arthritis (AIA), CIA, and K/BxN serum–transfer arthritis.

DTH was induced in WT and GILZ$^{″/″}$ mice with ovalbumin (OVA), and paw swelling was assessed at 24 hours using calipers as previously described. Lymph node T cells were stimulated with OVA (10–100 ìg/ml) for 24 hours, and interferon ã (IFNã) and interleukin 17A (IL 17A) enzyme linked immunospot (ELISpot) assays (BD Biosciences) were performed according to the manufacturer's protocol. DTH and AIA were induced in WT and GILZ$^{″/″}$ mice with methylated bovine serum albumin (mBSA)/Freund's complete adjuvant and assessed by measurement of joint swelling using calipers and histologic analysis, as previously described.

CIA was induced in DBA/1 mice, as previously described .Joints were scored daily for arthritis development as previously described, with a maximum possible score of 12 per mouse . Antigen specific T cell proliferation was analyzed as previously described ,using single cell suspensions of lymph node or spleen cells stimulated with concanavalin A (Sigma Aldrich) or mBSA. DNA synthesis (incorporation of ^{3}H thymidine [0.5 ìCi; GE Healthcare]) was measured using a liquid scintillation beta counter (Cambridge Scientific). K/BxN serum–transfer arthritis was induced as previously described, and each limb was scored daily from 0 to 5, allowing for the greater degree of swelling and erythema observed in this model. Histologic severity was evaluated on 4 ìm sections stained with Safranin O and counterstained with fast green/iron hematoxylin DBA/1 mice developing CIA were treated from day 21 with liposome encapsulated control and GILZ small interfering RNA (siRNA) sequences as previously described (11). Mice were injected intravenously with 200 ìl of cationic liposome/nucleic acid formulation 3 times a week. All animal experiments were performed in accordance with the regulations of the Monash University Animal Ethics Committee.

Lipopolysaccharide (LPS)–induced cytokine production.

WT and GILZ$^{″/″}$ mice were injected intraperitoneally (IP) with 10 mg/kg LPS. Mice were bled via the tail vein at 0 and 30 minutes and at 1, 2, 4, and 24 hours, and serum was collected. In some experiments 1 mg/kg dexamethasone (DEX) or vehicle was injected IP 1 hour prior to LPS injection. Murine dermal fibroblasts were isolated as previously described and maintained in complete RPMI 1640. Cells were seeded overnight and serum starved in 0.5% fetal bovine serum (FBS)/RPMI 1640 prior to 24 hours of treatment with 10$^{″7}$″10$^{″10}$M DEX with or without 200 ng/ml LPS.

RASFs.

RASFs were cultured as previously described. RA patients fulfilled the American College of Rheumatology 1987 revised classification criteria. All experiments were approved by the Human Research Ethics Committee, Monash Medical Centre. RASFs were transfected with 100 nM control siRNA or GILZ siRNA for 24 hours using RNAiMAX (Invitrogen) as previously described .Cells were treated with 1 ng/ml tumor necrosis factor (TNF; BioSource International) with or without DEX (10$^{″7}$″10$^{″10}$M), and supernatants were collected 7 hours later.

Enzyme linked immunosorbent assay (ELISA).

Anti–type II collagen (anti CII) IgG, IgG1, and IgG2a and cytokine concentrations were measured by ELISA as previously described .Murine TNF and IL 6 (R&D Systems), IL 1â (Endogen), and IFNã and monocyte chemotactic protein 1 (MCP 1) (BD Biosciences) were detected according to the recommendations of the manufacturers. RASF supernatant cytokine concentrations were analyzed using a cytometric bead array (BD Biosciences) as recommended by the manufacturer.

Construction, production, and administration of GILZ rAAV5.

GILZ rAAV5 was constructed as described .Briefly, the mouse gene coding for GILZ flanked by adeno associated virus type 2 (AAV2) inverted terminal repeats and under control of a cytomegalovirus promoter was packaged into the capsid from AAV5, resulting in the GILZ rAAV vector. Vectors were produced with an adenovirus free system in HEK 293 cells using a triple transfection method and purified by density gradient centrifugation. AAV titers were measured by quantitative PCR (qPCR), and final titers were 7 10^{13} vector genomes/ml (mock recombinant AAV [mock rAAV; empty vector]) and 3.85 10^{13} vector genomes/ ml (GILZ rAAV). DBA/1 mice with CIA (as above) were treated with an intraarticular injection of phosphate buffered saline (PBS), mock rAAV, or GILZ rAAV in both knee joints (5 ìl, 2 10^{11} vector genomes) and both ankle joints (2.5 ìl, 1 10^{11} vector genomes), administered in each mouse on the day of onset of arthritis (n = 10 per group).

Western blotting and qPCR.

Cell lysates were collected, and Western blotting was performed as previously described . Primary and secondary antibodies were rabbit anti GILZ (FL134; Santa Cruz Biotechnology), mouse anti–â actin (Sigma Aldrich), Alexa Fluor 680–conjugated anti rabbit IgG for GILZ (New England Biolabs), and Alexa Fluor 750–conjugated anti mouse IgG for â actin (New England Biolabs). Bands were detected using an Odyssey infrared imaging system (Li Cor). GILZ messenger RNA (mRNA) expression was measured by qPCR using methods and primers previously described.

RAW264.7 macrophage cell lines.

The P3K cell line derived from RAW264.7 macrophages expresses an NF êB luciferase reporter construct. Cells were maintained in á minimum essential medium (Gibco BRL) containing 10% FBS, 2 mM L glutamine, and 1% penicillin–streptomycin–amphotericin B. GILZ expression was induced using GILZ rAAV. After 72 hours cells were stimulated with 100 ng/ml RANKL for 6 hours, and luciferase activity was detected as described.

To further explore this, we next examined the effect of physiologic expression of GILZ on the K/BxN serum–transfer arthritis model, which is T cell independent but depends on chemokine mediated recruitment of myeloid effector cells. No significant difference in clinical or histologic arthritis severity was observed between WT and GILZ″/″ mice .Finally, we investigated the effect of physiologic expression of GILZ on LPS induction of TNF, IL 1â, and IL 6. LPS increased serum cytokine levels in both WT and GILZ″/″ mice, and no significant differences were observed. These observations were confirmed in vitro using thioglycollate induced peritoneal macrophages from WT and GILZ″/″ mice (data not shown). Taken together, these data suggest that despite its effects on T cell activation, physiologic expression of GILZ may not exert dramatic inhibitory effects on B cell activation, myeloid effector cell recruitment, or the expression of proinflammatory effector cytokines, and therefore not on models of RA that depend on these pathways.

Since the discovery that dimerization of the GR is not required for inhibition of the NF êB and AP 1 pathways (28), transrepression has been considered the dominant means through which glucocorticoids regulate inflammation. Recently, however, it has been shown that many effects of glucocorticoids on inflammation require homodimeric GR effects (29), and that selectivity for transrepression reduces effectiveness of GR ligands (30). This suggests that genes transcriptionally activated by glucocorticoids are essential for the control of immune–inflammatory responses. GILZ is among the most sensitively glucocorticoid induced proteins. Together with its ability to directly inhibit NF êB and AP 1, this suggests that GILZ is a potential mediator of the immune–inflammatory effects of glucocorticoids. It has not previously been established whether endogenous GILZ is required for the effects of glucocorticoids or whether it acts as a constitutive inhibitory regulator of the immune response like other glucocorticoid induced antiinflammatory proteins.

Our first experiments demonstrated increased OVA induced DTH responses and IFNã production in GILZ˝/˝ mice, consistent with previous findings in T cells of mice transgenic for GILZ. In addition, a previously unsuspected effect of GILZ on Th17 cell activation was suggested by increased IL 17A production in the setting of GILZ deficiency. The mechanism of these effects may depend on the reported effects of GILZ on Ras or on other pathways such as NF ATc1, as both affect T cell IL 17A expression . Unexpectedly, increased T cell activation in GILZ˝/˝ mice was not accompanied by increased AIA severity. Although AIA requires the activation of T cells , activation of effector cells resulting in the production of proinflammatory cytokines is also required, and such effects may vary between tissues. Our observations suggest that physiologic expression of GILZ is not a major regulator of effector responses to T cell activation, that increased T cell activation in the absence of GILZ was insufficient to exacerbate joint disease, or that GILZ deficiency resulted in a paradoxical insensitivity of local effector cells to T cell–driven activation.

To exclude the possibility that the lack of effect of endogenous GILZ on arthritis severity was specific to the relatively mild AIA model, we studied the more robust K/BxN serum–transfer arthritis model, which is mediated by chemokine dependent recruitment of myeloid cells . Here we also found no significant difference between WT and GILZ˝/˝ mice in clinical or histologic arthritis severity. Finally, LPS induced production of the cytokines TNF, IL 6, and IL 1â was unaffected by the absence of GILZ. Although multiple pathways are involved in the pathogenesis of arthritis models, one possible explanation for our findings is that the effects of physiologic expression of GILZ on T cells are not accompanied by effects on effector events such as antibody production, leukocyte recruitment, or cytokine expression that are essential for the development of arthritis.

Glucocorticoids act therapeutically via effects on both T cell and effector leukocyte function. The lack of effect of GILZ deficiency in these experiments called into question the hypothesized requirement for GILZ in the effects of glucocorticoids, suggested by previous in vitro studies using GILZ silencing . Therefore, we examined the requirement for GILZ in the effects of glucocorticoids, using GILZ˝/˝ mice and cells and in vivo siRNA silencing. In contrast to expectations, LPS induced serum cytokines were inhibited in vivo by DEX to an equivalent degree in WT and GILZ$^{-/-}$ mice, and dose dependent effects of DEX on cytokine release in vitro were equivalent in WT and GILZ$^{-/-}$ mouse cells. Moreover, the attenuation of CIA by DEX was unaffected by treatment with a GILZ siRNA regimen that we have previously shown to inhibit GILZ expression in vivo. Finally, GILZ silencing in RASFs did not impair DEX inhibition of cytokine production by these cells. Taken together, these data do not support the hypothesis that GILZ is required for glucocorticoids to exert their therapeutic effects on these aspects of the immune response.

Assessment of a potential therapeutic molecule based on results observed in its absence may be incompletely informative. Previous studies have demonstrated that T cell overexpression of GILZ leads to reductions in severity of experimental colitis and spinal cord injury in vivo . Therefore, to determine if GILZ could exert therapeutic effects in CIA despite the redundancy demonstrated by its deletion, we used rAAV as a means to induce the synthesis of GILZ in the joints . AAV serotype 5 has been shown to have tropism for dendritic cells, synovial fibroblasts, and endothelial cells in vitro, and to successfully deliver genes to the inflamed joint in vivo . Here, GILZ rAAV injection at the onset of disease in CIA successfully induced local GILZ expression, accompanied by significant attenuation of clinical and histologic joint inflammation. The reduction of joint disease in response to GILZ rAAV, despite the lack of effect on T cell proliferation and anti CII antibody production, is consistent with the restriction of GILZ overexpression to the joint and with the fact that T and B cell responses against CII would have been established prior to the time GILZ rAAV was administered, i.e., after disease was already evident. These effects also differ from those observed in AIA in the context of GILZ deficiency, in which increased T cell activation was observed without increased joint disease. These 2 observations, that systemic deletion of GILZ during the development of immune responses resulted in increased T cell activation but not arthritis, while local GILZ overexpression inhibited joint inflammation without affecting T cell responses, suggest differences between endogenous and supraphysiologic, and between local and systemic, effects of GILZ.

Despite the effects of exogenous GILZ, we demonstrated redundancy of endogenous GILZ with glucocorticoid antiinflammatory effects. No reported studies inform as to the potential redundancy of GILZ in the metabolic effects of glucocorticoids. However, GILZ has been described to promote osteoblast differentiation , suggesting the possibility that therapeutic GILZ induction may not transduce proresorptive signals of glucocorticoids. A wide array of studies is needed to probe the participation of GILZ in the metabolic effects of glucocorticoids, and the availability of GILZ deficient mice such as those we have generated here will enable such studies for the first time.

Several warnings apply to the interpretation of the results presented here. The mechanisms through which physiologic expression of GILZ affects T cell activation, and the functional consequences of these effects, require detailed examination. The lack of a phenotype in AIA in GILZ$^{-/-}$ mice is contradictory with the previously reported exacerbation of CIA in DBA/1 mice by GILZ silencing , which was confirmed by our current results with GILZ siRNA. Our GILZ$^{-/-}$ mice are on the C57BL/6 background, a strain that is relatively resistant to the induction of CIA .

Backcrossing the GILZ″/″ mouse strain onto the DBA/1 background would assist in addressing the effects of GILZ on CIA; however, the infertility of male GILZ″/″mice renders this infeasible. Moreover, it remains possible that a GILZ based therapy would be insufficiently broad in its effects to be efficacious in inflammatory disease. For example, the ability of GILZ to mimic the recently highlighted direct GR mediated repression of gene expression is unknown.

In conclusion, we have demonstrated the unexpected finding that despite attenuation of CIA in response to therapeutic GILZ overexpression, and increased T cell activation in GILZ″/″mice, GILZ is not a significant physiologic inhibitor of effector pathways of arthritis. Moreover, our results indicate that GILZ is redundant with the inhibitory effects of glucocorticoids on CIA and on LPS induced cytokine production. These findings suggest that, rather than being essential to the actions of glucocorticoids, the expression of GILZ may represent a "backup" pathway, wherein the effects of GILZ on transcription factors such as NF êB and AP 1 operate in parallel with direct GR transrepression of these pathways. Given the lethal effects of inflammation in the absence of endogenous glucocorticoids , the existence of such redundancy in glucocorticoid effects could confer a survival advantage. Nonetheless, the inhibitory effect observed in response to local GILZ overexpression in CIA suggests that further exploration of potential glucocorticoid mimicking therapeutic effects of GILZ is warranted. Understanding of the contribution of GILZ to metabolic effects of glucocorticoids would be required prior to advancing such an application to the clinic.

REGULATION OF THE ESTROUS CYCLE IN DOMESTIC ANIMALS

Neuroendocrine and endocrine factors involved in the regulation of reproductive cycles in domestic animals are discussed. Although research data from many species are considered, emphasis is placed on their relevance for the cow, sow, ewe and, to a lesser extent, the mare. Literature cited is not designed to be complete, but rather to be representative of the large volume of material which has been written on the subject.

Gonadotropin-releasing hormone is synthesized and secreted in response to various exteroceptive stimuli, but both its release and its effects on the anterior pituitary are modified by feedback of target gland hormones. A modulating role for nonsteroidal regulators such as inhibin has yet to be proven. Prostaglandins are important for corpus luteum regression and ovulation.

The relative roles of the gonadotropins and the above substances in hypothalamic, adenohypophyseal and ovarian function are considered. General mention is made of mechanisms of hormonal action.

Functional Causes of Infertility in Cows

Although problems in animal breeding are innumerable, infertility is one of the major bottlenecks in livestock production. In this paper functional causes of infertility like luteal deficiency due to abnormal pattern of LH pulses, cystic ovarian disease, and ovulatory defects, persistent corpus luteum, infertility associated with anestrus, non-detected estrus and repeat breeder are presented with respect to cow. It has been documented that in the management of cow, there are many times when the manipulation of normal cyclic activity ensures optimum production or is convenient for the owner by use of hormones to manipulate ovarian cyclic activity. Epitomizing under the prevailing condition increased productivity of the cow can only be achieved through success in minimizing infertility problems associated with functional cause through early diagnosis and treatment in addition to improvement of animal nutrition, health and genetic makeup.

Reproductive performance is one of the most important factors affecting dairy farm profitability and the development of national economy, as well as the living standard of rural and urban societies, Because, it directly or indirectly influences the yield of milk, reproductive culling rate and the cost for breeding and calf sales.. Dairy cows should calve one time every year to maximize economic efficiency. Cows that have been highly selected for milk production in recent decades have suffered a decline in cow fertility, fertility is a multi-factorial trait and its deterioration has been caused by a network of genetic, environmental and managerial factors and their complex interactions make it difficult to determine the exact reason for this decline.

The ovaries play the key roles in reproduction and any impairment in their functions can result in either sterility or infertility. The term fertility is applied to the cow denotes the desire and ability to mate, the capacity to conceive and to nourish the embryo and finally the power to expel a normal calf and fetal membrane. Sterility means an absolute inability to produce, whereas infertility denotes a degree of reduced fertility which results in failure to produce or delay in producing the normal live calf. Most forms of functional infertility results in anoestrus that means a failure of cows to display estrus.

Embryo survival is a major factor affecting production and economic efficiency in all systems of meat and milk production by ruminants . In beef cattle, estimates indicate that fertilization rate for oocytes is 90%, whereas average calving rates to a single service are between 40% and 55%, suggesting a rate of embryonic/ fetal mortality (excluding fertilization failure) of about 35% to 50% .. The majority of embryonic loss (70-80%) occurs in the first 3 weeks of pregnancy particularly between days 7 and 16 of pregnancy .

Hormones of reproduction are classified in to two groups according to their modes of action; primary hormones and secondary (metabolic) hormones. Primary hormones are involved in various aspects of reproduction, implantation and maintenance of gestation, parturition, lactation, maternal behavior and spermatogenesis. Primary hormones include FSH, LH and prolactin from anterior pituitary gland and oxytocin from posterior pituitary gland. Secondary hormones are necessary for the general wellbeing, metabolic state and growth of the cattle. Permit the full effect of the primary hormones indirectly related to reproduction. Secondary hormones are secreted by the pituitary, thyroid, parathyroid, adrenal cortex, uterus and pancreases. These hormones include Thyroid Stimulating Hormone, Adrenocorticotrophic Hormone, Anti-diuretic Hormone, Human Chorionic-Gonadotrophic Hormone and insulin.

Progesterone produced by the CL prevents cyclicity by acting on the anterior pituitary in a negative feedback fashion; therefore, decreasing the release of FSH and LH. It prepares the uterus for reception of fertilized ova and subsequent pregnancy. It also helps the cow maintain pregnancy by suppressing uterine contractions and promoting development of the uterine lining . prostaglandin $F_{2á}$ ($PGF_{2á}$) is secreted by the endometrial of the uterus and also affects structures on the ovary, helping to initiate ovulation by causing the demise of the CL, which results in withdrawal of progesterone's negative feedback mechanism. High circulating concentrations of progesterone stimulate the production and release of PGF2á from the uterus.

The postpartum period plays a pivotal role in Cow's reproduction. The duration of postpartum anoestrus has an important influence on reproductive performance. During the last few years, several studies have been attempted to treat the prolonged postpartum anestrum in cows by using hormonal treatments such as gonadotropin releasing hormone (GnRH), estrogen, prostaglandin ($PGF_{2á}$) and progesterone.

Follicular and Hormonal Dynamics during the Estrus Cycle

The estrus cycle is divided into different phases, (follicular phase, estrus, and luteal phase) and is regulated by hormones secreted by the hypothalamus (GnRH), anterior pituitary gland (follicle stimulating hormone (FSH) and luteinizing hormone (LH), ovary (estradiol and progesterone), and uterus (prostaglandin F2á (PGF2á). The preceding hormones serve as chemical messengers that travel in the blood to specific target tissues which contain receptors that are hormone specific and regulate the phases of the estrus cycle. The combination of hormone secretion and metabolism (liver, kidneys and lungs) maintain the correct hormonal balance during the follicular phase, estrus, and luteal phase of the cycle. A pre-ovulatory follicle and the subsequently formed corpus luteum are the two primary ovarian structures that regulate the estrus cycle through secretion of estradiol and progesterone, respectively.

Proestrus, the first phase of the estrus cycle which is building up phase, during this phase the ovarian follicle (under the influence of FSH and LH) enlarges and begins to secret estrogen, Starts in a day or two of regression of the corpus luteum from the previous cycle. Late in proestus the vaginal wall thickens and external genitalia will increase in vascularity in preparation for copulation.

Estrus, the period of sexual receptivity, is primarily initiated by the elevation in estrogens from mature follicles just prior to ovulation. In most cases ovulation occurs within a day or two after the onset of behavioral estrus.

Metaestrus, postovulatory phase dominated by corpus luteum function. During this period serum estrogen decreases and progesterone increases.

Functional Causes of Infertility in Cows

As a rule the functional causes of infertility tend to affect individual animals within a herd but in the aggregate they constitute an important cause of infertility; furthermore when they affect a large number of a particular sub group in a herd they frequently reflect some other problem, especially nutrition. Most functional aberrations occur because of some endocrinological abnormality which is frequently difficult to specify even with current methods of hormone assay, particularly when single, spot samples of blood or milk are examined. The abnormalities occur as a result of inherited factors; nutritional deficiencies or excess social influence which may arise from modern husbandry methods, for instance the grouping of large numbers of cows thus interfering with the establishment of a stable social hierarchy; and the stress of production.

Some of the common causes are elaborated as follows:

a) Non detected estrus(silent estrus)

b) Anoestrus

c) Ovulatory defects

d) Persistent corpus luteum

e) Cystic ovaries

f) Luteal Deficiency

g) Repeat Breeder

NON-DETECTED ESTRUS/SILENT ESTRUS

The cow show estrus but there is no one who can follow the cow. After the onset of puberty, cyclic ovarian activity should be maintained continuously through the cow's life except during pregnancy and for a short period in the puerperium. The only way that the herd's man knows that, this is occurring when the sign of estrus at approximately 21 days intervals. It is possible that there are signs but those they one not being observed; in this case it is a management problem.

The results of studies of the timing of estrus behavior were inconsistent with simple diurnal patterns and probability should be explained in terms of the periods of times for which cows were left undisturbed by other farm activities. Non-detected estrus can be confirmed by identification of cow's, provision of adequate lighting, heat mount detectors and other aids to the detection of mounting and daily follow up of the cow.

Anoestrus

Anoestrus is failure of cows to exhibit overt estrus but is more commonly a problem with estrus detection. Some cows exhibit overt estrus for a very short time, have few mounts or show signs of estrus in the middle of the night making estrus detection in these animals difficult. The estrus detection rate on many farms is less than 50%, being a very limiting factor to reproductive efficiency. Many factors, such as footing, management and milk production level, will affect demonstration of estrus. While these cows are not observed in estrus, they have normal estrous cycles and will respond well to ovulation synchronization programs. True anoestrus: It can be defined as the ovaries may be quiescent and inactive; this is referred to as a true anoestrus. The reasons for the failure of normal activity may be insufficient release or production of gonadotropins to cause folliculogenesis, or it may reflect the failure of the ovaries to respond . Clinical rectal palpation will reveal small ovaries which are usually flat and smooth, absence of corpus luteum, either developing, mature or regressing cow in true anestral will have virtually unchanged ovaries whilst a cow in late dioestrus or early dioestrus (metaestrus) will have a distinctly palpable corpus luteum . Equine chorionic gonadotropin hormone can be used to stimulate ovarian activity, it can induce follicular growth and estrus; at a dose rate of 3000-4500 IU. GnRH, Progesterone and estrogen has been used successfully to treat anoestrus in dairy cows.

Sub estrus or silent heat: There may be normal cyclic ovarian activity but the cow is not showing the normal behavioral signs; this is described as sub estrus or silent heat. The first and second ovulations post-partum is frequently not preceded by behavioral signs of estrus and is thus truly 'silent heat'. After the second estrus it is unlikely, that may result in true 'silent heats' to occur. When ovulation occurs in the absence of observed estrus it is more likely to be result of a failure of observation due to the short duration of behavior than to poor detection.

Diagnosis of the condition is made on the clinical history and rectal palpation of the genital system, no differentiation can be made from non-observed estrus, since the clinician will be checking for evidence of cyclic ovarian activity as demonstrated by a palpable corpus luteum. Determination of progesterone in milk or blood is a useful aid. In respect of treatment, if a mature corpus luteum is present and the cow is not pregnant, PGF2á analogue followed by fixed time insemination is indicated. If the corpus luteum is at refractory stage double injection of prostaglandin regimen at an 11 day interval be used. Alternatively other progesterone implant could be used followed by fixed time insemination.

Diagnosis of the condition is made on the clinical history and rectal palpation of the genital system, no differentiation can be made from non-observed estrus, since the clinician will be checking for evidence of cyclic ovarian activity as demonstrated by a palpable corpus luteum. Determination of progesterone in milk or blood is a useful aid. In respect of treatment, if a mature corpus luteum is present and the cow is not pregnant, PGF2á analogue followed by fixed time insemination is indicated. If the corpus luteum is at refractory stage double injection of prostaglandin regimen at an 11 day interval be used. Alternatively other progesterone implant could be used followed by fixed time insemination.

OVULATORY DEFECTS

Ovulation in the cow is a typical since it occurs 10-12 h after the end of behavioral estrus and 18-26 h after the ovulatory LH peak. During estrus and after the end of estrus several follicles undergo development but usually only one or occasionally two, ovulate; the other follicles regress and become atretic. The consequences for fertility of anovulatory defect are two-fold: either the oocyte is not liberated too late so that the spermatozoa are now incapable of fertilization, or else the oocyte has aged and is not capable of normal development. Ovulatory defects can occur due to two causes: endocrine deficiency or imbalance and mechanical factors. It is a complex interaction between ovarian hormones, the anterior pituitary and the hypothalamus. Thus if the quantity of pituitary hormone released is insufficient or its timing is incorrect (this is particularly true of LH), then ovulation is delayed or fails to occur.

Delayed ovulation: The hormone which is important for delayed ovulation is delayed LH surge which plays a big function for the failure of fertilization. If ovulation had not occurred by 24 h after service, the cow should be re-inseminated.

Diagnosis of the condition is difficult since it requires sequential rectal palpation of the ovaries which might self-intervene with of the ovulation and may cause premature rupture. It has been recommended that diagnosis can be made if the single follicle can be detected in the same ovary on two successive examinations, one at peak estrus and another 24-36 h later. Delayed ovulation is generally assumed to be one of the causes of failure of conception in cyclic non breeders . Treatment has consisted of using those hormones that might hasten the timing of ovulation. GnRH Administration causes a rapid rise FSH and LH concentrations peak within 30 to 60 min and return to pre injection values within 4 h. The pattern of release of

LH resembles the pre-ovulatory one observed during late estrus in normal cyclical animals..

Anovulation: Anovulation has been defined as the absence of ovulation which leads to both true anoestrus or to cystic ovarian disease. Sometimes anovulation is observed before the onset of a period of anovulatory anoestrus with follicle regressing and becoming atretic. If cows are examined per rectum during the first few weeks after calving, a number of enlarged ovulatory follicles can often be detected. They are incorrectly described as being cysts, but they are transient and don't persist even if no treatment is given.

Diagnosis of anovulation can only be made retrospectively by noting on rectal palpation or Ultrasonography that a follicle persists longer than one would have suspected .Treatment is directed towards ensuring that ovulation occurs at the next estrus, hence GnRH administered as described or delayed ovulation are indicated.

Persistent corpus luteum: Anything that interferes with the production or release of endogenous luteolysin, results in persistent corpus luteum. Pregnancy is the condition which most frequently results in the persistence of corpus luteum, but in the presence of uterine infection and inflammation of the tissues there is interference with the production or the release of luteolysin. There is little firm evidence that persistence of the corpus luteum can occur in the absence of uterine lesions.

Cystic ovaries: Ovarian dysfunctions like cysts occur most often during the early postpartum period when there is a transition from the non-cyclic condition during pregnancy to the establishment of regular cyclicity. It is generally accepted that cystic follicles develop due to a dysfunction of the hypothalamic pituitary-ovarian axis. This dysfunction has a multi factorial etiology, in which genetic, phenotypic and environmental factors are involved [26,27]. When discussing the pathogenesis of COF, a distinction may be made between a primary defect in the hypothalamus-pituitary and a primary defect at the level of the ovary in the follicle itself. However, COF formation may result from defects in ovary/follicle and the hypothalamus/pituitary as well.

The most widely accepted hypothesis explaining the formation of a cyst is that LH released from the hypothalamus-pituitary is altered: the pre-ovulatory LH-surge is either absent, insufficient in magnitude or occurs at the wrong time during dominant follicle maturation, which leads to cyst formation. It is believed that an altered feedback mechanism of estrogens on the hypothalamus- pituitary can result in an abnormal GnRH/LH release and cyst formation. A GnRH/ LH surge occurring prematurely during follicle growth, i.e., when no follicle capable of ovulation is present, can trigger the hypothalamus to be unresponsive to the feedback of oestradiol which results in the formation of ovarian cysts.

Follicular cyst: There are several definitions used to describe ovarian follicular cysts and the traditionally accepted definition is that they are "follicular structures of 2.5 cm or larger that persist for a variable period in the absence of a corpus luteum". We can consider ovarian follicular cysts to be any follicular structure on the ovary in the absence of luteal tissue, larger than normal follicular size that persists for a significant period of time and affects the estrus cycle of the animal.

Though we cannot determine the exact cause of ovarian follicular cysts, we can recognize that they "develop when one or more follicles fail to ovulate and subsequently do not regress but maintain growth and steroidogenesis" . It has also been determined that follicular cysts are anovulatory structures so, as long as they persist, cow Will remain infertile . Cystic ovarian disease is the "most common endocrine pathology to be found in dairy cows," and incidence is "believed to vary from 1 to 30% depending on herd and breed conditions". Holsteins are the most susceptible to develop a cystic condition compared to other breeds, and the most likely time of diagnosis is 30-60 days after parturition in high-yielding dairy cows.

Follicular cysts, when compared to other ovarian cystic conditions, are characterized by thin walls and produce very small amounts of progesterone. Occasionally, a persistent condition can lead to increased testosterone levels, causing some cows to exhibit masculine aggressive and sexual behavior. However, most cystic cows will remain in anoestrous as long as the condition persists. The fluid in follicular cysts contains many components, including hormones like estradiol, progesterone, and insulin.

One proposed explanation for the development of follicular cysts is that "the positive feedback of estradiol on release of gonadotropinreleasing hormone (GnRH) is compromised". This would not interfere with the ability of the pituitary gland to release luteinizing hormone (LH) in cystic animals, but the overall function of the hypothalamic pituitary- ovarian axis would be altered, and the preovulatory LH surge that normally induces ovulation does not occur. In cystic animals, despite the failure of ovulation by the pre-ovulatory follicle, a surge of follicle-stimulating hormone (FSH) would occur, and under conditions of low progesterone concentrations and high LH concentrations, dominant follicles would be caused to grow to a larger size. These oversized follicles are termed follicular cysts that produce high concentrations of estradiol and inhibin, cause a delay in follicular turnover and are responsible for the persistent cystic condition.

Luteal cyst: Luteal cystic ovarian disease is characterized by enlarged ovaries with one or more cysts, the walls of which are thicker than those of follicular cysts because of a lining of luteal tissue . Cysts with thicker walls produce high levels of progesterone. In appearance, they are smooth and rounded, with a spherical cavity that is lined by a layer of fibrous tissue surrounded by the luteinized cells "Luteal cysts are considered anovulatory cysts and are associated with infertility and mucometra in cattle". When compared to follicular cysts, luteal cysts are more likely to persist over long periods of time and can lead to nymphomania in some animals.

Luteal cysts "develop when ovulation fails to occur and the theca undergoes luteinization" . The luteal cyst occurs when the cells of the follicular cyst (granulosa and theca) become luteinized and start producing progesterone. Luteal cyst incidence increases with age and most often affects cows with high milk production.

Luteal deficiency: Progesterone is necessary for the maintenance of pregnancy until 150-200 days of pregnancy and perhaps in some cases to term. The main source of the hormone is the CL so that if this is not completely formed or it is not functioning adequately then insufficient progesterone is produced and the pregnancy fails. Luteal deficiencies have been suspected of causing infertility for many years and, although proof is difficult, cyclic non breeders or repeat breeders are frequently treated on this assumption.

It is impossible on rectal palpation to differentiate between a normal and abnormal CL; there is a natural variation in luteal size and the position of the CL within the ovary. It is variable and thus makes estimation of its size very difficult. By determining progesterone concentrations in blood or milk it has been possible to make some assessment of luteal function. Arthur et al. found that in 50% of cows that ovulated, yet failed to conceive, the plasma progesterone values six or more days after ovulation were lower than in the group that conceived. Extended low progesterone concentrations during the early luteal phase have been reported in 2% of cows and 18% of cows although some of the individuals which had this pattern conceived. In a study in 91 cow's post-partum the plasma progesterone values during the first dioestrus after calving were 34% lower in 10% of the cows than during the second cycle. In the same study the progesterone levels were similar in both pregnant and non-pregnant cows up to 11 days of the cycle.

Repeat breeder: The term repeat breeder or repeat breeder syndrome was created to describe cows that failed to conceive after, 3 or 4 inseminations in Cows those are cyclically normal with no clinical abnormalities. Repeat breeder females return to service repeatedly after being bred with a fertile male. The factors which are responsible for these abnormalities are multiple, but the major ones are anatomical, hormonal, managemental and infectious and vary from herd to herd and animal to animals.. Repeat breeder cow's exhibit normal signs of estrus every 18 to 24 days but require more than 3 services to become pregnant.

In practice, some will have been inseminated at the wrong time, others may have pathological changes in the bursa or oviduct that are difficult to palpate or undiagnosed uterine infections.

Signs of estrus every 18 to 24 days but require more than 3 services to become pregnant Some of these cases may be associated with early embryonic deaths, failure of fertilization, poor semen quality and most of the embryonic losses in cows occur much earlier in pregnancy than previously believed. Diagnosis of repeat breeder relies on the animal records, rectal palpation, ultrasonography, and vaginal examination.

Ovulation is initiated by the increase of the luteinizing hormone (LH), which results in the rupture of the follicle and release of an egg. After ovulation there is a short period during which the egg can be fertilized. Since the period between the first signs of estrus and ovulation occurs at 72 h or later, the cows that ovulate later should be inseminated daily until rectal detection of ovulation is confirmed. Delayed ovulation should not be confused with repeat breeder syndrome (repeat breeders). These cows, if not inseminated daily until ovulation, can also return to service repeatedly after artificial insemination while showing normal signs of estrus every 18 to 24 days.

Treatment of Different Signs of Endocrine Disturbances

Cystic ovarian diseases like follicular cyst is treated with hCG at a dose 3000 to 4500 IU, GnRH at a dose of 100-250 ìg for luteinization, at a dose of 0.5 ìg for ovulation. GnRH applications more than 80% of the cows come in to estrus within 18 to 24 days. For luteal cyst use Progesterone at a dose of 100 mg in oil intra-muscularly on three consecutive occasions at 48 h interval. $PGF_{2á}$ and its analogue at a dose of 25 mg.

Anoestrus is treated with PMSG at a dose of 3000 to 4500 IU to stimulate the ovarian activities and induce follicular growth and ovulation, GnRH at a dose of 0.5 mg to stimulate LH release. But in suckled cows second injection after 10 days is important after transient raise in progesterone to initiate normal cycle.

Sub-estrus/silent heat is treated with $PGF_{2á}$ If there is matured CL, one dose of PGF2á can stimulate LH secretion, If there is no matured CL, double administration of $PGF_{2á}$ the first dose for maturation of CL and the second for progesterone release. GnRH is the best treatment for ovulatory defects and also it improves pregnancy rate, because it can raise FSH and LH rapidly.

Uterine lavage with a PGF2á application and without an antibiotic application may be a preferable treatment method for repeat breeder cows. A gonadotropin-releasing hormone (GnRH) application significantly increases the conception rate in repeat breeders. Therefore, GnRH must be considered in the hormone treatment for repeat breeders.

CONCLUSION AND RECOMMENDATIONS

Animal production contributes much to the development of the national economy and to the better living standards of the rural and urban area as well. Reproductive performance is the one that can influence the profitability of the dairy farms in terms of increasing or decreasing milk yield, reproductive culling rate and market price of dairy cows. It can be affected by a network of genetic (hereditary), environmental and managerial factors (malnutrition, lack of intensive follow ups and stress) and their complex interactions.

Functional causes of infertility are secondary to the basic nutritional, hereditary, stress and work factors. As we know the functional causes of infertility usually affect individual animals within a herd but in the aggregate they constitute an important cause of infertility; but in a wide range effect, when this problem affects a large number of a particular group in a herd they frequently reflect some other problem, especially nutritional problems.

Fertility is a multi-factorial character and its deterioration has been caused by those factors listed above. In order to increase the production and productivity, adequate data are required with respect to the functional and also other causes of infertility in cows.

Since the past few years, In order to treat those reproductive aberrations, hormonal treatments and managemental advancements were some of the corrections given by the veterinarians and farm owners to improve the reproductive performance of their animals.

Based on the above conclusive statements, the following recommendations are forwarded:

- Adequate diagnostic facilities for reproductive problems are needed.
- There should be a means for frequent and detailed examination of reproductive organs.
- Awareness among farmers regarding management of reproductive disorders due to nutrition leading to imbalance or inadequate hormonal levels should be created.

3

Cattle Breed

Over 800 breeds of cattle are recognized worldwide, some of which adapted to the local climate, others which were bred by humans for specialized uses. Breeds fall into two main types, regarded as either two closely related species, or two subspecies of one species. Bos indicus (or Bos taurus indicus) cattle, also called zebu, are adapted to hot climates. Bos taurus (or Bos taurus taurus) are the typical cattle of Europe, north-eastern Asia, and parts of Africa - they are referred to in this list as "taurine" cattle, and many are adapted to cooler climates. Taurus/indicus hybrids are widely bred in many warmer regions, combining characteristics of both the ancestral types (such as the Sanga cattle of Africa). In some parts of the world further species of cattle are found (both as wild and domesticated animals), and some of these are related so closely to taurine and indicus cattle that interspecies hybrids have been bred. Examples include the Dwarf Lulu cattle of the mountains of Nepal with yak blood, the Beefalo of North America with bison genes, the Selembu breed of India and Bhutan with gayal genes. The Madura breed of Indonesia may have banteng in its parentage. The Dzo of Nepal is an infertile cattle-yak crossing which is bred for agricultural work. Like the mule which is also infertile, they have to be continually bred from the parent species.

Cattle were originally used as triple purpose animals. They were used for draft, milk and meat. As horses, then machinery began to replace them for work purposes; breeds were developed for dual and single purposes.

How are cattle classified?

The main divisions of cattle are based on primary use:

Dairy (milk)

Beef (meat)

How are the physical characteristics of cattle classified?

The selection of physical characteristics of cattle is based on their use. Dairy cattle should match an ideal dairy type, while selection for beef cattle will be for a beef-type animal. A good dairy cow will be rectangular in shape with good body capacity and strong legs. Udder conformation is especially critical in the milking breeds. A beef cow will be stockier with more emphasis placed on muscling. There is more variation between breeds among beef cattle.

Do different breeds of cattle have variations in personality, behavior and type?

There is quite a bit of variation in behavior and type between breeds. Some breeds are known for their calm disposition, others for their excellent mothering ability, others for their hardiness…there is no perfect breed. Selection will depend on a great number of factors including; environmental conditions, the primary use of the animal, type of forage available and of course, personal preference.

Which breed information should be learned for testing purposes?

In addition to being able to identify (photos) the breeds listed below, the following information will be required (1) origin (2) primary colors and/or patterns (3) at least one distinctive characteristic related to its disposition, behavior or function. Use Internet links (http://www.ansi.okstate.edu/breeds/cattle/) and books to find the answers to your questions and to study the breeds.

DAIRY BREEDS

- Ayrshire
- Brown Swiss
- Guernsey
- Jersey
- Holstein-Friesian

The Infertile Mare

Sometimes it can be quite difficult to get a mare pregnant, especially if she has fertility problems. There are several reasons that a mare might be infertile, many of which can be treated. Two reproductive specialists from Lexington, Ky., shared their extensive knowledge of several of the problems that mares can face with attendees of the Thoroughbred International Exposition and Conference.

Wolfsdorf began her presentation with the three natural physical barriers of the mare's anatomy that help prevent contamination of the reproductive tract. The vulvar labia is the lip-like outer portion of the mare's genitalia. The vulvovaginal fold, part of the the hymen, is a fold of mucous membranes found at the opening of the vagina. And finally, the cervix is the narrow opening at the beginning of the uterus. Compromise or failure of these barriers can cause particulate matter or air to enter (pneumovagina) to enter the reproductive tract. The resulting contamination could result in cervicitis (cervical inflammation), vaginitis (vaginal inflammation), or endometritis (inflammation or infection of the endometrium or uterine lining).

Conformation problems of the reproductive tract or changes over time, possibly as a result of multiple foals, can result in retention of urine in the vagina. This is called vesicovaginal reflux, or more commonly urine pooling. This problem results in reduced conception rates, changes in lumenal pH, periglandular fibrosis (formation of scar tissue), or inflammation.

Wolfsdorf said that mares which have had numerous foals usually develop poor reproductive conformation because of the loss of muscle tone and damage to the vulvar suspensory ligament, retractor muscle, and the vulvar constructor muscle. She said it is best to examine older mares when they are in heat since the effects of estrogen will increase the vascularity and relaxation of the muscles, giving the veterinarian a better picture of what problems might have to be dealt with before breeding.

Other less common causes of mares pooling urine an ectopic ureter, cystitis, and urolithiasis. An ectopic ureter is one in which the ureter (the tube that conveys urine from the kidney to the bladder) opens somewhere other than the bladder. Cystitis is inflammation of the urinary bladder, and urolithiasis is the presence of calculi, or mineral deposits, within the urinary tract.

Diagnosis of reproductive problems, specifically urine poolers, involves a reproductive exam, which can include a speculum exam (best done during estrus), uterine culture and cytology to check for inflammation and contamination, a biopsy for long-term prognosis, transrectal ultrasound for uterine fluid, and enzyme tests to check the fluid for changes in pH, creatinine levels, and urea.

Treatment will depend on the findings of the exam and various tests. For urine pooling, if the problem is mild and is caught early, then the urine can be removed and the horse's physical condition can be improved to strengthen the muscle tone. In more severe cases, medical management is needed. This might include uterine lavage (flushing), use of the medications oxytocin or bethanecol, which might improve urethral sphincter tone (the opening of the urethra into the vagina), acupuncture, or surgery to extend the urethral sphincter closer to the opening of the vulva.

Another problem that Wolfsdorf said might cause infertility is fungal endometritis, which occurs when a fungus or yeast contaminates the uterus and causes infection and inflammation. She said this problem is hard to treat and has a high rate of recurrence. Causes can include poor vulvar conformation, repeated and/or lengthy intrauterine antimicrobial treatment, and excessive manipulation of the reproductive tract. The most common culprits are the *Candida* spp. and the *Apergillus* spp. Diagnosis of fungal endometritis can be confirmed with uterine cytology and culture, ultrasonography which reveals highly particulate fluid, and the presence of vulvar/vaginal discharge.

Treatment can involve uterine lavage with large volumes of fluid, the use of dimethyl sulfoxide (DMSO), diluted betadine or vinegar, or antifungal agents such as clotrimazole, amphotericin, fluconazole, and nystatin. Wolfsdorf said that the disadvantage to antifungal agents is that they are expensive and the duration that treatment is necessary is unknown. Despite treatment, prognosis might still be poor since treatment only affects certain stages of fungal development, the organism might have attached to the endometrial folds, or the mare might have delayed uterine clearance. The chance for re-infection is high as it is hard to destroy or flush out all of the organism.

LeBlanc focused her part of the presentation on problems of older mares. She said that 30% of mares over the age of 14 will need veterinary assistance to conceive. One problem older mares face is persistent follicles, or cystitic follicles. These follicles do not ovulate in response to the normal signal from the brain. These follicles will show up on an ultrasound as a black area with white strands, which indicates blood clots. (Normal follicles show up as all black areas on an ultrasound.)

LeBlanc recommends that breeders continue to tease the mare until she goes out of heat. Five to six days after she is out of heat, her progesterone levels can be measured. If they are between one and four ng/ml, then prostaglandin might help. If the progesterone level is less than one ng/ml, then there is no luteal tissues surrounding the cystic follicle and all one can do is wait for the mare to come back into estrus on her own. LeBlanc said these follicles do not respond to human chorionic gonadotropin (HCG) or the hormonal drug Ovuplant.

Human chorionic gonadotropin should only be used on follicles that are at least 35 mm, for the best response, since the follicle must be "sufficiently mature" before it will respond to HCG. It has been found that as mares get older, some do not respond to HCG.

LeBlanc explained that uterine inflammation is a normal process after breeding. The mare's body treats sperm as a foreign object it should attack. After the attack, immune cells then engulf the sperm, and try to clear it from the body. All of this happens after 15-30 minutes, so any sperm that has not reached the oviduct during that time will be under attack.

Another problem of older mares is accumulation of fluid in the uterus. This could be due to decreased cervical drainage as the uterus loses the ability to contract properly, or due to decreased lymphatic drainage (which removes particulate matter and fluid swelling). If treatment is started after too much fluid has accumulated, it's hard to get rid of, said LeBlanc. Oxytocin can be used to produce uterine contractions for 20-40 minutes, four to six hours apart, which helps clear the fluid.

The prostaglandin cloprostenol (Estrumate) can produce contractions for up to five hours for clearance of fluid and a return to normal of the lymphatic system. LeBlanc warns against using cloprostenol longer than 12 hours after ovulation.

What is Infertility? Both the mare and the stallion can experience reproductive issues. Lack of estrous cycle in the mare and low production of sperm in the stallion are just two of the obstacles that may be seen in the breeding process. Calling in your veterinarian to evaluate both breeding partners may be necessary in order to determine where the infertility problem originates. Hormonal therapy and antibiotics in case of infection may be utilized to encourage a successful mating. Infertility in horses is defined as the inability of a horse to conceive. Many factors can affect the ability of your mare or stallion to produce offspring. Symptoms of Infertility in Horses Stallions – He may be sub fertile; he only manages to impregnate mares 50% of the time or less Small testicles Evidence of back pain Signs of arthritis Infection Low sperm count Mares - Failure to conceive in a breeding season; your mare may not conceive during a breeding season at all despite all efforts made Lack of willingness to mate Evidence of infection such as vulvar discharge Signs of injury Causes of Infertility in Horses The possible causes of infertility vary greatly for both stallions and mares. Mares Endometritis – Typically the most common cause of infertility in mares; it is an inflammation of the uterine lining Infection – A large problem with infections are that they can go without being noticed for significant periods of time and are not easy to treat Inflammation - Ovarian hormones have a large part in inflammation of the uterus and can cause fertility problems as well Scarring/trauma – Injury to your horse's reproductive organs can make conception difficult or impossible as well Hormonal – Abnormal cycles resulting in difficulty for the mare to conceive or carry to term Neoplasia – Tumors on your horse's ovaries can cause her to struggle with fertility Stallions Low sperm count – This could be hereditary or directly related to your horse's reproduction organs Testicular degeneration – Over time or even via trauma, your horse's testes may not function as well as they once did resulting in lowered fertility Ejaculation issues – This could be a partial ejaculation concern or could be the side effect of other concerns/issues STDs – While relatively uncommon in horses, these diseases can result in fertility issues while your horse is infected and afterwards Trauma – Stallions can be injured by mares during breeding and this results in ongoing trauma to their penis; this can make mating very difficult as well Tumors – These can result in infertility for your stallion either via injury or sickness Diagnosis of Infertility in Horses Diagnosis of infertility will be done via a physical exam of your horse by his or her veterinarian. The physical exam will allow your veterinarian to find any obvious causes of infertility such as twisted testes in your stallion, or injury to your horse's reproductive organs in the mare. Once the physical exam is done, imaging may be necessary as well. One imaging test used is the ultrasound. Ultrasounds allow your veterinarian to get a better look at your horse's internal organs to determine possible causes for the infertility. An endoscopy of the urethra of the stallion may be suggested. Biopsies can be taken from your mare's uterus and if your stallion has any tumors those may be biopsied as well. Tests may also be run on semen, uterine cultures or blood to determine underlying causes of infertility. In some cases, there can be more that one problem, encountered concurrently, that is causing the infertility of either horse. Treatment of Infertility in Horses Treatment of your horse's infertility will largely be dependent on the underlying causes. In the event of STDs, infections and hormone imbalances for example, medication management may be an option to get these underlying issues controlled. If endometritis is the concern, there are options such as a routine of douching your mare's uterus with a salt solution that can result in increased fertility. Cleansing of the penis of the stallion in the case of sheath irritation, NSAIDs due to injury, and surgery in the case of tumors may be required. Regulation of the estrous cycle and sexual rest for the stallion are other treatment methods that may be employed. Recovery of Infertility in Horses Ongoing follow up will be necessary as directed by your veterinarian. Recovery time will be dependent on what, if any procedures were done. Changes may be necessary to accommodate aging horses to assist them in mounting when breeding for example. Most concerns are not life threatening to your horse and many of the fertility issues can be treated accordingly.

Endometritis, a common cause of infertility in mares

Infertility, resulting from failure to conceive during the preceding breeding season, was caused by various forms of endometritis in forty-four mares of Highland and Arabian breeds in Yugoslavia. Chronic mucopurulent and latent catarrhal endometritis occurred most frequently. Douching of the uterus on alternate days for 9 days with warm and cold sterile salt solutions (ranging from 7% to 1% w/v) in the early autumn resulted in conception in 47-7% of mares at an average of 15-2 days after the onset of treatment, and another 36-7% becoming pregnant in the following spring. Mating at the wrong stage of the oestrous cycle and too frequent mating during oestrus were judged to be the most common causes of endometritis in the mares examined

MARE INFERTILITY

What is normal fertility? The following are the rules of thumb regarding expectations at a well-run breeding farm using natural cover or artificial insemination.

1. A 50% pregnancy rate is considered normal after a group of mares are bred through their first estrus or heat cycle (single estrus).
2. An 80-85% pregnancy rate is considered a normal range in a group of mares bred during a breeding season.
3. A pregnancy loss of 10-12% on well-managed horse farm is not abnormal.
4. If you have 100 mares, the average will be 80-85% pregnant at 40 days and 70-75% will produce a live foal.

5. Early embryonic loss is a major factor in fertility losses in sub-fertile mares. A 60-70% loss prior to day 14 can be expected.
6. Early embryonic loss on an average group of mares is considered to be 10 to 15 %. These are detectable losses generally after 14 days of pregnancy.

IRCCS SDN, NAPOLI, ITALY

1. Introduction

Every year many mares fail to become pregnant. These failures represent a substantial economic and genetic loss to the horse industry. Most major infertility problems are complex and several factors, singly or in combination, can cause failure to produce offspring. In the last few years, much of the efforts of practitioners and researchers working in equine breeding industry have been directed to individuate the pathophysiological mechanisms underlying poor reproductive performances in mares. To clarify, infertility includes three "problems" mares' types: mares that fail to cycle, mares that cycle normally but do not conceive and mares that cycle normally and conceive but then suffer early embryonic death. Endometritis is on the talk in much of the recent research as the most frequent cause of the last two conditions.

Post-breeding persistent endometritis, bacterial and other infective endometritis and poor uterine clearance have all been discussed in an attempt to define risk factors and a diagnostic algorithm, essential to reach a definitive diagnosis and to apply the appropriate therapeutic protocol. Breeding-induced endometritis is a normal physiological reaction in the horse, as it is believed that an inflammatory response is necessary for the effective removal of contaminating bacteria and excess spermatozoa introduced into the uterus. In a healthy uterus, the inflammation subsides within 48 hours, but the susceptible mares are not capable of resolving the inflammation triggered by sperm and develop persistent mating-induced endometritis (PMIE). A very strong relationship establishes between PMIE and infectious endometritis and a complex of mare (such as age, perineal conformation, uterine clearance and cervical competence) and microbial (such as induction of inflammation, epithelial adherence, resistance to phagocytosis and viscosity of secretion) factors contribute to the pathogenesis of endometritis. Traditionally, the mares that are prone to endometritis are called susceptible, in contrast to the "resistant" ones, not prone to uterine infection.

The aim of this chapter is to perform a thorough review of recent literature about endometritis. The cascades of inflammatory signals being complex and intertwined, the etiopathogenetical, diagnostic and prognostic roles of the recently studied inflammatory markers are discussed. In addition, the most common bacterial and fungal pathogens involved are reviewed, together with the recent advances in diagnostic procedures. In fact, the diagnostic algorithms are carefully examined, highlighting pros as well as pitfalls of each diagnostic aid. Suggested therapeutic protocols are examined in the effort to detect what is actually recommended and what would better benefit from further corroboration, with special attention to the correct use of antimicrobials and antibiotics, their common way of administration and contraindications. Consideration will be given to therapy alternatives such as proper breeding management, use of uterine lavage, oxytocin/prostaglandin administration and treatment of the biofilm formation. The idea that a better etiopathogenetical understanding of the endometritis remains the key to access to a correct diagnostic protocol and to a successful therapeutic plan will inspire this chapter.

2. Pathophysiology of endometritis in the mare

Endometritis is a major cause of infertility in the mare. It is an acute or chronic inflammation of the endometrium that can be classified based on both its etiology and pathophysiology. Susceptibility to persistent bacterial endometritis was characterized as early as 1969 . Since then, the physiopathological mechanisms involved in persistent endometritis and its correlation with bacterial endometritis and infertility in the mare have been discussed in several studies [2–18]. Endometrium responds to the introduction of air, urine, semen, bacteria, fungi or yeasts through an inflammatory reaction that ultimately hesitates in restoring an environment suitable to receive the conceptus, as it migrates from the oviduct few days after the insemination. To induce endometritis in experimental conditions, either spermatozoa or bacteria such as *Streptococcus zooepidemicus* or *Escherichia coli* are commonly infused into the uterus. A subpopulation of mares, designated as susceptible, fails to resolve the endometritis and develops a persistent inflammatory condition, which affects fertility. Furthermore, if the mare is unable to clear bacteria that may have entered the uterus during breeding, a bacterial infection can develop [19, 20].

In conclusion, any difficulty in the physical clearance of inflammatory debris from the uterus after mating or foaling triggers the endometritis. Although the mechanisms for uterine clearance for different antigens are closely related, pathophysiology, clinical signs and therapy partially differ. For these reasons, breeding-induced endometritis and bacterial endometritis will be described separately.

Transient and persistent mating-induced endometritis

The transient breeding-induced endometritis is a physiological reaction in the immediate hours after breeding. It is a local inflammatory response necessary to remove excess spermatozoa and bacteria introduced into the uterus . This response is limited to local inflammation, because any hematological alteration in inflammatory parameters has not yet been detected during breeding-induced endometritis . Furthermore, in healthy mares, the endometritis resolves within 24–48 hours, leaving the uterus clean and free from inflammation.

Mechanical clearance and inflammatory response

Semen and its extender play an important role in the induction of defense mechanisms, depending on seminal components and sperms numbers, concentration, viability and site of semen deposition. In mares mated, or artificially inseminated with fresh or cooled semen, seminal plasma activates the complement system that evokes massive migration of the polymorphonuclear leucocytes (PMNs), cytokines and mononuclear cells invasion that prepares the endometrium to recipe the embryo. During the cryopreservation process, an important mechanism of modulation of the inflammatory response is lost due to the removal of seminal plasma, so that a severe inflammation follows the insemination with frozen/thawed semen. Furthermore, seminal plasma contributes to the transport and survival of viable spermatozoa and the elimination of non-viable spermatozoa from the uterus, suppressing binding between neutrophils and viable spermatozoa .

The complement activation cascade led to the formation of leukotriene B4, prostaglandin (PG) E and PGF2± and other arachidonic acid metabolites, which act as chemo-attractants for PMNs in the uterus. The PMNs, on their part, drive the inflammatory response to cleaning the uterus: spermatozoa and bacteria destruction is performed by way of phagocytosis and release of neutrophil extracellular traps, composed of extensions of DNA and histones with antimicrobial action . The intrauterine fluid is composed of neutrophils, inflammatory mediators and plasma proteins, including immunoglobulin and enzymes. The release of PGF2± stimulates myometrial contractions combined with ciliary propulsion of the mucus blanket, promoting the elimination of bacteria and dead inflammatory cells via the cervix or the lymphatics.

In "resistant" mares, uterine fluid is eliminated and inflammation resolves within 5 days post-insemination and the fertilized oocyte descends in a uterus ready to implantation . On the contrary, susceptible mares are unable to resolve the physiological breeding-induced inflammation and they surrender to persistent post-insemination endometritis. The accumulation of intrauterine fluid persists over five days post-ovulation in susceptible mares . Therefore, the success of getting the mare pregnant is compromised by an unprepared endometrium and by the development of concomitant bacterial infections.

It is likely that at the origin of both breeding-induced and bacterial endometritis, there are common failures in the physical mechanisms of uterine clearance by uterine contractions. Indeed, mechanical clearance plays a key role in the uterine response to the contamination, and deficits in myometrial contractility could have a major responsibility in the pathogenesis of delayed uterine clearance. The increase in myoelectric activity, indicating an increase in uterine contractions, observed in healthy mares after the insemination, resulted delayed by 2 hours in susceptible mares. Additionally, susceptible mares showed a sharp decline in activity, dropping below baseline levels after 12 hours . Furthermore, it was observed that more radiocolloid was retained in the uterus of susceptible mares 2 hours after infusion than in resistant mares . Impaired myometrial function could be related to the increased intrauterine accumulation of nitric oxide (NO), produced in excess by inducible NO synthase (iNOS) during inflammation in susceptible mares, as confirmed by an increased endometrial expression of iNOS mRNA after insemination. The NO is a smooth muscle relaxant and myometrial tissue was unable to respond to electrical stimulus in the presence of NO, during in vitro experiments.

Reaction to the insemination could involve humoral and cellular mechanism, with differences between resistant and susceptible mares. Several studies produced contradictory findings, but most authors agree that immunoglobulins are involved mainly in the endometrial response to bacteria than to semen. However, there are no data to demonstrate that susceptibility is a direct cause of altered immunoglobulins in the uterus. Similarly, it was suggested that in resistant and susceptible mares, different inflammatory cells were involved and that susceptibility to persistent endometritis was primarily caused by a dysfunction in the adaptive immune response. Until now, there are no data to prove these hypotheses, and just a decreased opsonizing ability of the uterine secretions in sub-fertile mares compared with normal mares was demonstrated .

The inflammatory response in endometritis is a complex process involving multiple signaling pathways, initiated by the cytokines, produced by a variety of cell types, to allow the recognition of antigens and the recruitment of inflammatory cells . Specifically, the inflammatory response is modulated by a delicate balance between the expressions of pro- and anti-inflammatory interleukin (IL). Another important pro-inflammatory cytokine involved with the inflammatory response is interferon-. It promotes the migration of inflammatory cells through vessel walls and leads to the activation of microbicidal functions, and to an upregulation of iNOS . Few studies focused on these inflammatory pathways in the mare so far, and they gave contradictory results in relation to the experimental protocol and the timing of evaluation. Specifically, timing is very important in the experimental design, since the difference between resistant and susceptible mares consist mainly in the persistence of the breeding-induced inflammatory response. Thus, results of the studies were sensibly different, measuring the expression of several IL at 24 hours or at 6 hours after breeding. However, an increased mRNA expression of IL8 and lower expression of IL10 were observed in susceptible mares compared to resistant mares 24 hours after insemination . On measuring cytokines expression in mares at several time points within the first 24 hours after breeding, susceptible mares had lower expression of the inflammatory modulating cytokines IL10, IL1-receptor antagonist and IL6 when compared with resistant mares 6 hours after insemination. After 3–6 hours from breeding, there were no differences between resistant and susceptible mares in the degree of uterine inflammation, fluid retention, endometrial cytology and ultrasound imaging. However, on the basis of the differences in mRNA expression of cytokines observed, it was stated that a critical time in the development of persistent breeding-induced endometritis occurs around 6 hours after breeding .

Predisposing factors to persistent breeding-induced endometritis

Persistent endometritis has multifactorial pathogenesis, and properties of the bacteria and mare's characteristics are key components in the development and resolution of the inflammation . Several predisposing factors are associated with the individual mare. Young and fertile resistant mares and older and subfertile susceptible mares have been examined to clarify the differences in the ability to resolve uterine inflammation. Indeed, the aging and parity have been associated with altered systemic immune response. Older mares present several predisposing factors to develop persistent uterine infections after breeding. Aging and parity influx on the ability of clearing debris, since mares develop over time anatomical or degenerative defects that interfere with uterine drainage. Other risk factors associated with persistent infections include the conformation or the acquired alteration of the internal and external reproductive organs. For example, not only poor vulvar conformation, incompetent vagino-vestibular sphincter, vaginal stretching or an incompetent cervix but also genital pathology, such as pneumovagina or vagino/

cervical injuries can facilitate the entrance of pathogens, which normally lives over body surface. Mares with cervical fibrosis secondary to a traumatic birth, or with the elongated, narrow cervix typical in aged maiden mares, accumulate uterine fluid easily because of the compromised cervical drainage . Free fluid accumulation is also facilitated by a pendulous uterus or impaired lymphatic drainage and atrophy of endometrial folds .

Degenerative changes, such as an abnormal myometrium, periglandular fibrosis, vascular degeneration, lymphangiectasia, scarring and atrophy of endometrial folds or damage to the reproductive apparatus may also explain delayed uterine clearance of bacteria, fluid and debris . The altered mucociliary activity does not impede the bacterial adhesion on the endometrium and expulsion of the inflammatory cells. Vascular degenerations inhibit hormones delivery to endometrium and disturb uterine drainage, reducing venous return in capillary beds. Prolonged endometrial edema and consequent persistent inflammation characterize the uterus of susceptible mares.

Bacterial endometritis

The sexually transmissible disease in horses are caused by primary pathogens Taylorella equigenitalis, certain unspecified serotypes of Pseudomonas aeruginosa and Klebsiella pneumoniae capsule types 1, 2 and 5 . The true venereal disease is caused by T. equigenitalis and is known as contagious equine metritis resulting in cervicitis, vaginitis and endometritis. Asymptomatic infected carrier stallions transmit the pathogens to the mares, which will present copious muco-purulent vaginal discharge within a week after breeding. However, this pathology is endemic in Europe and a more insidious form with minimal clinical signs has been recognized. P. aeruginosa and K. pneumoniae inhabit the external genital of the stallion and infections are transmitted by coitus, insemination with infected semen and genital manipulations .

Contamination of the uterus by fecal and genital opportunistic flora may provoke an infection during mating or genital tract manipulations. A wide variety of opportunistic aerobic and anaerobic bacteria, fungi and yeasts, either alone or in synergy, have been occasionally implicated as causes of endometritis. In a report, Streptococcus equi subsp. zooepidemicus was responsible for approximately 65% of the cases while Escherichia coli, K. pneumoniae and P. aeruginosa accounted for approximately 10%. The alteration of uterine defense mechanisms is attributed to individual microbial factors such as induction of inflammation, epithelial adherence, resistance to phagocytosis and viscosity of secretions . Bacterial products can modify any properties of mucus, rendering cilia unable to expel uterine exudates. Changes in production, viscosity or elasticity of mucus and cilia function determinate the adverse effects on uterine clearance and hence, these can interfere with antibiotic penetration, resulting in treatment failure or antibiotic resistance .

Chronic endometritis

Chronic infection and mixed population of microorganism cause severe, progressive and irreversible fibrotic condition that affects mare endometrium. Long-standing influx of lymphocytes and plasmacells into the endometrium contribute to chronic degenerative changes, such as periglandular, perivascular or diffuse stromal fibrosis. This condition impairs endometrial function and future pregnancies, causing infertility.

Diagnosis of endometritis

The diagnosis of endometritis in the mare represents one of the crucial challenges for the equine clinical practice . Equine practitioners have recently understood that the major cause of the progression from an acute to a chronic condition is the failure to identify the causative agent, given that affected mares do not always show evident clinical signs . Hence, it is critical to make a correct etiological diagnosis, and to characterize the degree and the type of inflammation, in order to establish an effective treatment. Furthermore, it is not uncommon that mares with a long history of normal fertility can acquire post-breeding endometritis. In such cases, the clinician has no opportunity for prophylactic intervention .

The diagnostic algorithm for the identification of the causative agent comprises a detailed reproductive history and a complete clinical evaluation, including the transrectal palpation of the reproductive tract and the use of ancillary diagnostic aids. Among the latters, specific instrumental investigations, such as transrectal ultrasonography, vaginal and cervical exams, with both manual and endoscopic evaluation, and laboratory tools, such as uterine culture, endometrial cytology and biopsy, represent useful tools for the diagnosis of endometritis. Recently, myeloperoxidase has been investigated as a uterine inflammatory biomarker, but the relationship between its concentration and pathologic uterine conditions needs further studies . To increase the pregnancy rates in "problem" mares, the practitioner should dispose a quick and reliable diagnostic technique to start the best treatment as soon as possible in the breeding season .

EXPERIMENTAL IDENTIFICATION OF SUSCEPTIBLE MARES

Mares with endometritis begin to show intrauterine fluid accumulation following breeding, after the first three to four successful pregnancies. Uterine degenerative changes appear and worsen with age, the reproductive tract may become more pendulous and uterotonic drugs are no longer effective . Furthermore, some reproductively normal, nulliparous as well as pluriparous mares exhibit an excessive uterine inflammatory response after being bred with frozen semen. The exuberant reaction may be a consequence of a low volume of seminal plasma and the lack of its beneficial action in decreasing the migration of neutrophils into the uterine lumen .

The delay in uterine clearance has an important role in the pathogenesis of uterine inflammation, since it influences the retention of endometrial inflammatory factors, whether they are bacterial or sperm-induced . In fact, it was shown that mares susceptible to chronic uterine infections failed to clear adequately Cr-labeled microspheres when compared with mares with normal uteri. The uterus, studied with scintigraphy in mares with a delayed uterine clearance, showed to be oriented vertically, while in reproductively normal mares it was oriented horizontally. Hence the "baggy" uterus contributes to fluid accumulation and low clearance. However, although scintigraphy has been proven to be a useful method for studying post-breeding endometritis, it seems not to be a functional diagnostic tool in a clinical setting. In addition to the uterine position, the myometrial function plays an important role in uterine clearance, as confirmed by electromyographic recordings . Nevertheless, not all mares susceptible to endometritis exhibit a delayed uterine clearance if cervical dilatation is appropriate .

Clinical identification of susceptible mares

History

Mares susceptible to developing a persistent post-breeding endometritis are more often pluriparous, barren, older than 12 years of age and with a poor perineal conformation. Nonetheless, some maiden mares, which show an insufficient degree of relaxation of their cervix during estrus, may also develop persistent mating-induced endometritis, regardless of age. In such mares, the problem may be resolved with the first foal delivery .

Clinical examination

The mares' reproductive tract examination should include a body condition scoring and the notation of previous or existing foot problems . Any changes in perineal, vulvar or cervical conformation should be evaluated before breeding. The two vulvar labia may not create a proper seal due to persistent relaxation of the vulva, and the rectum may be displaced cranially, stretching the vulva dorsally to the ischiatic arch and causing repeated fecal contamination of vulva and vestibule . This kind of conformation is linked to the loss of fat in the perineal area due to a high level of fitness, and muscular fatigue and estrus exacerbate the perineal relaxation, but it can also be a heritable trait. It is very important to check that the cervix opens properly in estrus and closes in diestrus .

Clinical signs

Clinical signs of endometritis may be hidden, but vaginal discharge, short inter-estrus intervals and/or a shortened luteal phase and reduced fertility can be detected. In bacterial endometritis, they can vary widely according to the pathogen, with symptoms markedly different between Gram-positive and Gram-negative bacteria . A vaginal discharge may be seen more often during estrus . However, the presence of evident clinical signs 48 hours post-breeding is of concern, since a non-inflammatory uterine environment is important for embryonic survival .

IMAGING OF THE REPRODUCTIVE SYSTEM: ULTRASONOGRAPHY

Ultrasonography (US) has always been considered an accurate and essential step in the diagnostic workup of endometritis. The monitoring function of US is as important as the diagnostic one, since mares with subclinical endometritis require careful daily checking post-breeding . Since its inception, the use of US has strongly impacted on the ability to detect the presence and to quantify the volume of uterine fluid accumulation, which is considered the main sign of endometritis, especially when observed after breeding; but this "hallmark" sign is not always present. Intrauterine fluid presence has also been shown to be strongly related with low pregnancy rates, especially in mares older than 16 years, and to the biopsy score, revealing its diagnostic value in identifying the uterine inflammation. It may be detected as soon as 6 to 36 hours after breeding, but it can be noted at the pregnancy examination 14 to 16 days after breeding. Nonetheless, the more significant time to detect the uterine fluid collection is mid-to-late diestrus because the cervix is closed, and in a healthy uterus, fluid should be little or absent . Intrauterine fluid has been also correlated to specific endometrial bacteria and cytological findings, with mares positive to non-pathogenic bacteria showing intra-uterine fluid in <40% of the US examinations. The relevance of the uterine fluid accumulation can be evaluated by examining the depth, the echo-texture and persistence of the fluid . Two or more centimeters of intra-uterine fluid during estrus, or between 6 and 36 hours post-breeding, are significant indices of susceptibility to mating-induced endometritis. A high echogenicity in the uterine fluid is associated with the presence of neutrophils and debris, thus indicative of inflammatory components . Short, thick and hyperechoic lines within the uterus may represent air or exudates.

Sub-Fertile Mares

A broodmare which fails to conceive or carry a foal to term can be an emotional and financial liability. For the backyard breeder, the liability might be primarily emotional, when that special mare you wanted to produce your next trail horse or.

A broodmare which fails to conceive or carry a foal to term can be an emotional and financial liability. For the backyard breeder, the liability might be primarily emotional, when that special mare you wanted to produce your next trail horse or hunter simply doesn't do so. For the commercial breeder, there might be emotional trauma, but more importantly for him or her, the economic liability can be significant.

It costs an estimated $4,000-$5,000 per year to maintain a broodmare at a top facility in a place like Kentucky. That is not small change for a mare which has, as her prime function in life, the bearing of offspring for commercial sale or private athletic use.

There can be a multitude of reasons why a mare doesn't conceive or carry a foal to term, and some of them have to be laid at man's doorstep. Mares in the wild which don't bear young travel with the band until the end of their days, but they don't leave behind a legacy of more problems through their offspring. It is Nature's way of culling, just as it is when a horse has weak hooves and falls prey to a predator. Only the fittest survive to produce replacements.

Man has not been content to accept Nature's edict, mostly for economic reasons. If a mare has been a superior performer on the racetrack or in the show ring, for instance, the owner's goal when the mare has retired from competition, or even before, is to raise a foal that will be of equal or greater ability. The fact that the mare tells us through her physiological response that she is nearly incapable of reproducing doesn't deter us. We turn to science to manipulate her system to achieve a pregnancy that will be carried to term.

Science has provided many answers and tools in recent years, but with these positives come some negatives. Yes, we often can obtain a foal, but we might be continuing an inherent reproductive weakness that will be passed on from generation to generation, thereby challenging future researchers even more.

That being said, there is little doubt that man will continue in his quest to outwit Nature in the reproductive battle. In today's society, that is perhaps as it should be and will result in some stellar equine athletes that otherwise would not have been born.

CAUSES OF SUB-FERTILITY

Sub-fertility is multi-faceted. If one were to attempt to group or categorize the reasons, they would fall into three general areas: 1) age, 2) contamination and infection of the reproductive tract, and 3) a malfunctioning of the hormonal system. The three will overlap, but for the purposes of this discussion, we'll approach sub-fertility from those viewpoints.

The information that follows comes from a variety of sources, including papers presented at past gatherings of the American Association of Equine Practitioners.

One of the basic reasons mares become sub-fertile is age. Old mares often have a more difficult time conceiving and carrying a foal to term than do young mares. With some aged mares, the problem might involve simple wear and tear on a uterus that has become scarred and compromised to the point that it can no longer adequately house and nourish a growing fetus.

In others, as research at the University of Wisconsin has indicated, the problem might occur before the fertilized egg ever reaches the uterus. The Wisconsin study indicated that ovarian failure was associated with advanced age–mares 20 years and older.

The researchers found that more young mares than old mares had at least three ovulations in a period of approximately 90 days during the natural breeding season from June through September.

Elaine M. Carnevale, DVM, MS, PhD, formerly at the University of Wisconsin and now at Southern Illinois University, who authored a paper summarizing the Wisconsin research, had this to say:

"In mares with at least two sequential ovulations, significant differences included later emergence of the ovulatory wave of follicles, slower growth of the ovulatory follicle, longer interovulatory interval, longer follicular phase, and less luteinizing hormone surge in old than young mares. An atypical appearance of the ovulatory site was imaged with ultrasonography more in old than young mares and might have been associated with luteinized, unruptured follicles or incompletely evacuated follicles. Reproductive senescence in old mares progressed over a number of years as follows: a) elongation of the follicular phase and a reduced number of follicles when compared with other mares, b) occasional ovulations–80 days between ovulations in some cases–and, c) a final state of persistent ovarian inactivity with no follicles larger than five millimeters."

The Wisconsin researchers carried out a second study that was designed to determine whether uterine changes and deterioration in older mares had a negative impact on pregnancy.

Carnevale had this to say

"In another study, mares 15 years of age and older had significantly more inflammatory and degenerative changes of the uterus and less uterine contractile activity than young mares. These results agreed with previous conclusions that functional and histologic changes occur in the uterus with advanced age. However, to determine the extent that fertility was reduced in old mares prior to the time that the embryo enters the uterus (Day 5 or 6) oviductal embryos from young and old (20 years of age and older) mares were evaluated.

"By Day 1.5, the collection rate of ova (cleaved or uncleaved) and the percentage of cleaved ova were significantly lower in old mares compared to young mares.

"On Day 3, significantly fewer embryos were recovered from old mares, and embryos from old mares had more morphologic abnormalities and retarded growth. Oviductal pregnancy rates on Day 3 were comparable with pregnancy rates of Day 11 for young and old mares, respectively. Results indicated that fertility was reduced in old mares prior to the time that the embryo entered the uterus.

"Oocytes, collected from young and old donors, were transferred into the oviducts of young, artificially inseminated recipients. Significantly fewer oocytes collected from old mares, when compared with young mares, developed to an embryonic vesicle by Day 12. Oocyte viability was reduced in old mares."

The conclusion reached by the Wisconsin researchers was that fertility problems in older mares often occurred earlier in the reproductive cycle than many horse owners had thought.

"Although uterine changes were present in old mares, fertility was reduced prior to the time that the embryo entered the uterus. Therefore, uterine pathology did not appear to be the primary reason for low fertility in old mares. Oocytes were defective and resulted in a significant loss of fertility in old mares. Investigations are needed to determine if defects are intrinsic to oocytes of old mares or if they result from a detrimental follicular milieu and whether oocytes could be salvaged from a harmful environment or treatments could be developed to enhance the viability of follicular oocytes."

Carnevale also was involved in research funded by Purina Mills. The study was aimed at determining whether diet could have an effect on ovulation and pregnancy rates among older mares.

Young mares (three to five years of age) and old mares (19 and older) were used in the study. The mares were primarily of Thoroughbred or Standardbred lineage. They were housed on a 12-acre lot and fed hay. On March 30 of that year, young and old mares were randomly divided into two groups. One group would receive a "nutrient-dense" Purina feed that had been designed specifically for old mares. It would be fed at the rate of 10 pounds per day in addition to alfalfa/orchard grass hay. The second group would receive oats at the rate of 10 pounds per day, plus the same kind and amount of hay.

Reproductive tracts were examined every one to four days via ultrasonography.

The results indicated that the young mares fed the special diet did not differ significantly in ovulation from young mares on the oats-hay diet. However, there were distinct differences in the two groups of older mares.

Carnevale had this to say in her report:

"During the spring, old mares ovulated smaller follicles and had more multiple ovulations than young mares. Old mares fed a highly digestible, nutrient-dense diet ovulated approximately two weeks earlier in the spring than old mares fed an isocaloric control.

"Type of feed did not affect time of ovulation for younger mares. Old mares fed the special diet ovulated on approximately the same mean date as young mares, in contrast to a later ovulation date for old mares fed the control diet. Differences in effects of diet between young and old mares were probably caused by the reduced capabilities of old mares to digest and assimilate feed, resulting in a functional nutrient deficiency. From these data, it is apparent that older horses need to be maintained on a feeding program that supplies them with the additional nutrients that they need for reproductive performance."

CERVICAL LACERATIONS AND OLD MARES

Before leaving the discussion of older mares and the reproductive problems attributed to age, it would be well to note that older mares which have had a foal regularly are more prone to cervical lacerations, even when the delivery is normal.

Of course, a mare of any age can suffer a cervical tear during a difficult birth (dystocia), but age appears to be a definite factor, even when there are no birthing problems.

Corey D. Miller, DVM, of Texas A&M University, conducted a study on the subject, using the medical records of all mares referred to Rood and Riddle Equine Hospital in Kentucky for cervical lacerations between March of 1986 and January of 1995.

His findings were twofold. First, as has been noted, age was a factor in cervical lacerations and, second, surgery to repair the damage often was successful. The post-surgical pregnancy rate for mares in the study was 75.3%. However, he said, clients should be made aware of the potential for recurrence with subsequent foalings.

Following is Miller's conclusion

"The results of this study suggest that older (mean age, 12.8 years), multiparous (mean number of foals produced before surgery, 6.2) mares may be predisposed to cervical lacerations. Eighty-six percent of the mares in this study population were presented for surgical laceration repair after an apparently normal parturition. This supports previous anecdotal information that suggested that cervical lacerations are just as likely to occur during an unremarkable parturition as during a dystocia.

"Surgical repair is an effective treatment of cervical lacerations. The potential benefits of the surgery appear to far exceed the few complications associated with the procedure."

Infections

The above studies fairly well establish that age is a significant factor in the reproduction process. But, what about the younger mares which don't reproduce? What might be involved with them and even some of the older mares which are ovulating regularly? Infection somewhere along the reproductive tract can be a significant factor.

Each portion of the reproductive tract plays a key role in the mare producing a live foal, from ovaries to fallopian tubes and from vulva to uterus. Any time a part of the reproductive tract becomes contaminated or infected, its ability to function properly could be compromised.

Unfortunately, it is difficult to keep the mare's reproductive system free of contamination. For example, if the vulva is tipped in, fecal matter will fall against it every time the mare defecates. This can result in the reproductive tract becoming contaminated or infected. Or, the labia might lose elasticity and be unable to seal tightly, allowing airborne contaminants to make their way into the reproductive tract.

The way in which the uterus functions when the mare is in heat as opposed to diestrus also can be a factor in contamination. When the mare is not in estrus, the cervix is tightly closed and located in the center of the vaginal wall. When the mare is in heat, the cervix is open and relaxed, actually lying on the vaginal floor. During natural cover breeding, the stallion's penis might penetrate the uterus through the cervix. The same is true of instruments used in artificial insemination. Whatever the approach, the door is open for contaminants.

Contamination of the reproductive tract can result in endometritis, an inflammation of the endometrium, the inner layer of the uterus that is responsible for nourishment of the fetus. Mares which are susceptible to endometritis lack the ability to clear the uterus of bacterial contamination that has been created by breeding.

Mares susceptible to endometritis often will require 96 hours or more to clear the uterus of contaminants, while reproductively normal mares will clear the bacterial contamination within 24 to 48 hours.

"Uterine inflammation that occurs because of breeding must be cleared by 96 hours if the embryo is to survive," says Michelle M. LeBlanc, DVM, Diplomate ACT, of the College of Veterinary Medicine, University of Florida.

"Recent research completed by ourselves and others," she wrote in a paper on the subject, "has shown that a delay in the physical clearance of intrauterine fluids and bacteria through the cervix is a major cause of recurrent endometritis. We have also shown that lymphatic drainage of the uterus is impaired in susceptible mares.

"Lymphatics remove excess fluids that accumulate in tissue spaces and carry proteins and large particulate matter away from tissue spaces, neither of which can be removed by absorption directly into blood capillaries. The reason why susceptible mares exhibit a delay in uterine clearance is not known. Contributing factors include poor perineal conformation, anatomical location of the uterus in relation to the pelvic brim (my preliminary findings), and decreased myometrial contractions after antigenic stimulation."

LeBlanc said that mares having a delay in uterine clearance can be diagnosed by a new scintigraphic procedure performed during estrus or within 48 hours of ovulation. Radiocolloid (radioisotopes in pure form) is infused into the uterus and the amount of colloid cleared in two hours is measured.

Reproductively normal mares, says LeBlanc, clear 50% of the colloids in that time, whereas susceptible mares clear only negligible amounts. The only problem with the method is that the scintigraphy equipment is expensive and is not readily available to the practitioner in the field. Perhaps a more practical method, says LeBlanc, is an ultrasound examination of bred mares between one and 10 days after ovulation to measure intrauterine fluid.

LeBlanc theorized that treatments designed to improve myometrial contractions or physical clearance could be useful in older mares susceptible to endometritis. Her research goal was to determine the effects of oxytocin on uterine clearance in the laboratory and in clinical cases.

Seven reproductively sound mares and seven mares susceptible to endometritis were used. Mares declared to be reproductively normal were either maiden mares or mares which had produced a number of foals and had no history of infertility.

Clearance of radiocolloid was measured with scintigraphy in all mares on Day 3 of estrus and 48 hours after ovulation. Radiocolloid was infused into the uterus and the initial scintigraphy reading was taken. Then, 20 IUs of oxytocin were given intravenously and clearance of colloid was measured over two hours.

LeBlanc had this to say about the results:

"Oxytocin enhanced the uterine clearance of radiocolloid infused into the uterine lumen in mares resistant and susceptible to endometritis during estrus. Administration of 20 IU of oxytocin intravenously to susceptible and resistant mares resulted in clearance of 90% and more radiocolloid within 15 minutes of intrauterine infusion. The susceptible mares cleared negligible amounts of radiocolloid (about 5%) within four hours when no drug was given.

"Following oxytocin administration, a few mares had sweat on their flanks and shoulders. No other adverse effects were noted."

A second study carried out at the University of Florida involved field trials conducted during the 1993 and 1994 breeding seasons.

Oxytocin was administered in combination with saline uterine lavage to 32 infertile mares. Treatments were performed four to eight hours after breeding and were repeated 24 hours after breeding if all three liters of saline infused into the uterus were cloudy when retrieved.

Oxytocin, also at the 20 IU level, was given intravenously after the uterine lavage was conducted. The mares, ranging in age from four to 24, were stationed at various farms, but all of them had histories of two or more years of recurrent endometritis.

Of the 32 mares involved in the treatment program, reported LeBlanc, 27 (84%) conceived after mating during one estrus.

In her conclusion, LeBlanc had both positive and precautionary remarks about the use of oxytocin:

"Oxytocin dramatically enhanced the clearance of radiocolloid in reproductively normal mares and mares susceptible to endometritis. Oxytocin, in combination with a saline uterine lavage, also improved pregnancy rates in infertile mares exhibiting a delay in uterine clearance. By using the combined treatment, we removed intrauterine fluid containing inflammatory by-products by saline lavage, and uterine contractions are induced with oxytocin. It is not known whether oxytocin by itself will improve pregnancy rates.

"Oxytocin must not be considered a wonder drug because not all mares with endometritis exhibit a delay in uterine clearance. Veterinarians must continue to examine diligently all mares referred for reproductive consultation. Conformational defects have to be noted and corrected surgically if necessary. Examination of the vagina and uterus must be performed as aseptically as possible to reduce iatrogenic infections. Ultrasonographic examination of the reproductive tract should be conducted, if equipment is available, to determine whether fluid is accumulating within the uterine lumen."

OXYTOCIN AND ANTIBIOTIC LAVAGE

A study conducted at the University of Utrecht, The Netherlands, and reported on by J. F. Pycock, BVM, PhD, DESM, MRCVS, added another dimension to using oxytocin–antibiotics were administered as well. Approximately 380 mares–some maiden, some barren, and some which had foals–at seven well-managed public stud farms in the United Kingdom were examined. The ages of the mares ranged from three to 22 years, and the examinations took place during the period of February through September, 1993.

Once in estrus, the mares were placed in one of four treatment groups: 1) no treatment; 2) broad-spectrum, water soluble, intrauterine antibiotic treatment; 3) intravenous injection of 25 IU oxytocin, and 4) an intravenous injection of 25 IU oxytocin followed by intrauterine antibiotic infusion.

Of the 95 mares given only antibiotics, 54 (57%) became pregnant. Of the 95 given oxytocin alone, 62 (65%) became pregnant. Of the 95 given antibiotics and oxytocin, 73 (77%) became pregnant. The lowest pregnancy rate was for the non-treated mares (less than 60%).

"In non-treated mares," reported Pycock, "more fluid accumulated in the uterine lumen after mating, which was the most likely reason for the significantly reduced pregnancy rates."

The results of this study have shown the benefits of treatments that either cause the elimination or that reduce the production of intrauterine fluid in the management of post-mating endometritis in the mare. The study avoided the use of an endometritis model and utilized large numbers of mares in normal clinical practice. An important advance was the inclusion of untreated control mares.

"Drainage of fluid following oxytocin injection usually was rapid and visible on ultrasound examination. No untoward clinical signs were observed in any of the mares receiving oxytocin. Pregnancy rates were highest in the mares given intrauterine antibiotics as well as oxytocin. This must suggest two different modes of action of the combination treatment, namely antibacterial activity and fluid drainage, which were additively helpful.

"Antibacterial activity may possibly be linked to the reduction of the intraluminal fluid, because in those mares that were treated with oxytocin alone, some bacteria remained within the uterine lumen, and presumably, the inflammatory effects of these bacteria caused the fluid to accumulate again.

"From the results, it is necessary for one to ask the question, 'Is it desirable to treat every mare routinely after mating regardless of the clinical history?' The answer must be no, because treatment should be restricted to those mares that would benefit. It may be useful to use an ultrasonographic examination of the uterus to identify those mares with uterine fluid after mating and to treat only those mares.

"Another approach that appeared useful was to target categories of mares that benefitted the most from treatment in terms of pregnancy rate. This study indicated two such categories–old mares and mares mated at the first estrus postpartum. This may offer the best approach to prevent unnecessary use of postmating therapy."

Reproductive Conformation

Conformational abnormalities often are implicated in reproductive tract infections, some as the result of the aging process and others that are traced to genetics.

We already have mentioned the mare with the tipped-in vulva, seen with some frequency in horses with high, level croups. In cases that result from genetics, one can do nothing to change the mare's basic conformation. This leaves the veterinarian with the option of suturing the upper section of the vulva leaving only an opening for urine to be eliminated–a Caslick's procedure. The purpose is to prevent particulates from entering the reproductive tract from post-mating to foaling time.

The procedure is quite straightforward and simple and is used on many mares where conformation creates the suspicion of contamination.

Not quite so straightforward and simple is the mare which has conformation that allows urine to pool in her reproductive tract, also known as vesicovaginal reflux (VVR) or urovagina. This is a serious problem along the reproductive tract that often is implicated in sub-fertility. In some cases, the only remedy is surgery.

Simply put, urine pooling occurs when urine refluxes in the vagina and against the cervix instead of being flushed to the rear and evacuated through the vulva. The problem is magnified during estrus because of the relaxed state of the cervix, which allows urine to enter the uterus, irritating and contaminating it.

Urine in the forward portion of the reproductive tract, as well as in the uterus itself, produces chemical irritation as well as spermicidal activity. The irritation it causes can provide a breeding ground for harmful bacteria.

Again, age becomes a pre-determining factor. Older mares which have delivered a number of foals are more at risk than younger females, says Gayle Trotter, DVM, a surgeon at Colorado State University who has written textbook literature on the problem and the surgical procedures involved in correcting it.

When the problem does occur in younger mares, says Jack Easley, DVM, MS, of Shelbyville, Ky., also the author of textbook literature on VVR, it could mean urine pooling with added complications. The young mare might have an ectopic ureter–displacement of the tube that conveys urine from the kidney to the bladder–and, as a result, might empty urine into any area along the reproductive tract. This is a problem that can only be solved with surgery.

Under normal circumstances, urine is evacuated from the bladder into the vulvar vestibule and out the open vulvar labia or lips. However, if the normal slope of the vulva is altered approximately 10 degrees from horizontal or the genitourinary tract slopes forward and downward, a splashback of urine can occur, whereby some urine refluxes back toward the cervix and uterus instead of out of the vulva.

The problem is multifaceted. Mares which are wind-suckers, for example, are more apt to be urine poolers. Wind-sucking occurs when the vulvar labia become weakened and do not function properly by providing a tight outer seal for the reproductive tract. When that occurs, air is drawn into the tract, inflating the vaginal vault.

Mares which are wind-suckers (pneumovagina) are prime candidates for urine pooling because when she draws air into her reproductive tract, the vaginal vault is inflated and the entire tract becomes relaxed, eliminating normal barriers.

One of the reasons that barriers along the tract might not be working well, says Trotter, is the bad conformation that has developed, often from multiple births. The loss of muscle tone allows the vagina to tilt forward and one of the barriers, the sphincter, is unable to provide a tight seal in the tract.

There is another side to the coin. Some mares will urine pool only when in estrus. This is the least serious of the urine pooling problems, and the one most easily treated.

Easley explains it this way

"Depending on the stage of estrus and concentration of estrogen and progesterone, the reproductive tract may be predisposed to urine pooling. Estrogens stimulate and maintain the tissue of the tubular reproductive tract by increasing vascularity and uterine secretions, therefore increasing the area's defense against mechanical insult and infection. A side effect, however, is a mild edema of the tubular tract, vagina, and uterus, resulting from intracellular absorption of water. The pelvic structures and vulva then relax as the general perineal area enlarges and cranial displacement of the reproductive tract follows, leading to urine pooling."

Often, says Trotter, these mares can be managed successfully by manually evacuating the urine, then breeding them. When a pregnancy occurs, the mare's hormonal changes will cause the reproductive tract to return to its normal, firm state and permit the cervix to seal off the uterus.

Because the urine pooling problem might only be a mild one that can be solved without surgery, it is important that the correct diagnosis be made. A surgical procedure on a mare which pools urine only during estrus might do more harm than good.

Trotter, however, adds a note of caution involving mares which only pool urine during estrus:

"Some of these mares become pregnant if the urine is manually evacuated prior to breeding. However, some of these mares also likely suffer early embryonic death of the fetus or accumulate urine more significantly in subsequent years. A full reproductive workup, including a uterine biopsy for prognostic purposes, is recommended prior to embarking on a course of medical or surgical treatment."

For the wind-sucking mare that is also a urine pooler, a Caslick's operation might be the treatment of choice.

When urine pooling persists for whatever reason, however, an internal surgical procedure could be the only solution.

Endometritis

To this point, we have discussed some ways to prevent endometritis, plus some of the basic culprits along the way. When endometritis does strike, it must be dealt with immediately by the practitioner in the field if the mare is to be a viable producer.

"It has always been amazing to me," says Jim Briddle, DVM, a Kansas State University graduate and practitioner near Riverton, Wyo., "that we don't get more uterine infections post-coitus. After all, the mare's cervix relaxes completely, and when the stallion breeds her, his penis goes through the cervix and all the way into the uterus. Thus, there is great potential for bacteria to be deposited in the uterus."

Listed by many veterinarians and researchers as the most common bacteria involved in endometritis is Streptococcus zooepidemicus. Two other prime culprits are Pseudomonas aeruginosa and Klebsiella pneumoniae.

The good news at this stage is that while a multitude of bacteria enters the uterus during the breeding procedure, this is also the time when the mare's immune system is functioning at its peak. Thus, most healthy mares are able to fight off the bacteria. However, as already indicated, a number of mares are unable to do this and must be treated.

In order for the practitioner to know what he or she is dealing with specifically, a culture is necessary, says Briddle. Only when the veterinarian knows exactly what bacterial organism is involved can the correct decision be made as to which antibiotic is appropriate.

"A practitioner can do a lot with a microscope right there at the clinic," Briddle says. "If there are bacteria present, you are going to be able to see them and you may be able to determine what general type is involved."

By studying a uterine culture in the clinic, Briddle feels the practitioner can get a head start on treatment while awaiting official results from a laboratory. One of the most beneficial services the laboratory provides, he says, is to isolate the bacteria and recommend an antibiotic that sensitivity tests have demonstrated has the capability of destroying it.

One of the procedures the practitioner can perform while awaiting laboratory results is cleansing the uterus via uterine lavage with one to two liters of sterile saline, plus, as mentioned above, administration of oxytocin.

Uterine Cysts

Repeated bouts with endometritis can result in uterine cysts, says Scott Bennett, DVM, of Equine Services, a clinic and equine hospital in Simpsonville, Ky., that is heavily involved in the reproductive field. The cysts will develop in an older mare after multiple births. They are rare in young mares unless those mares have had ongoing bouts with endometritis.

The cyst problem crops up in the wake of scar tissue that develops within the uterus as the result of infection or multiple births. What happens then is a clogging of the lymphatic draining system by the scar tissue. The result is a cyst or cysts filled with fluid that normally would be evacuated.

"A few cysts within the uterus won't make that much difference, unless they are large," says Bennett. "Their ability to compromise the function of the uterus is dependent on their size and number. Normally, I don't worry about two or three small cysts, but one big cyst can become a major problem.

"In some mares, there will be 10 or 12 cysts, and their presence in that number will definitely cause problems even though they may not be large. The cysts become physically obstructive, sometimes preventing semen from entering the uterus and, if a pregnancy does occur, can compromise mobility of the embryo. That is how the embryo gives the signal that a pregnancy has occurred, by wandering through the uterus for the first 18 days."

When cysts are suspected, it is time for a thorough uterine examination, says Bennett.

Following are some of the tools and approaches modern technology has made available to the practitioner to conduct a complete uterine exam.

Endometrial culture–An endometrial culture is a general evaluation of the bacterial status of the uterine lumen. "It must be remembered," says Bennett, "that a negative culture simply means that you did not culture any bacterial growth from that swab. Conversely, a positive culture does not necessarily mean that the uterus has a pathological condition and is subject to interpretation."

Subfertility

A successful breeding program begins with the careful management of your broodmares. Mares need to be evaluated prior to each breeding season to screen for infections or any other conditions that may affect fertility. Mistakes made in the management of your mares will often lead to increased veterinary costs down the road. Broodmares should enter the breeding season healthy and in good condition. This means they should have a Body Condition Score of 5-6.5. Broodmares should have their teeth floated, their vaccinations should be current, and they should be maintained on a regular deworming schedule (or fecal testing protocol).

Sometimes mares have difficulty conceiving. There are a variety of reasons for impaired fertility in mares. These are simply a few of the many causes of infertility in mares:

Endometritis
Poor sperm quality
Impaired uterine clearance
Post-breeding metritis
Poor vulvular conformation
Hemorrhagic anovulatory follicles
Endocrine dysfunction
Persistent CL's
Poor oocyte quality (older mares)
Oviductal problems
Endometrial cysts
Systemic illness
Ovarian tumors
Incompetent cervix

Screening / Pre- Breeding Exam

This exam is performed prior to breeding a mare and will include an ultrasound examination of the reproductive tract. This is done to screen for any abnormalities on the ovaries or in the uterus and to determine the stage of the reproductive cycle the mare is in. Potential fertility affecting conditions that would be found on an ultrasound examination can include ovarian tumors such as a Granulosa Cell Tumor or Uterine Cysts. Both of these conditions can be treated and in doing so the fertility of the mare can be repaired. A uterine culture and cytology will also be performed. This test provides information as to the quality of the uterine environment and will detect uterine infections that can cause infertility. These infections if present are identified and sensitivities are performed in our laboratory to determine the precise antibiotic required to effectively treat the infection. Uterine lavages are performed to "wash" the uterus to reduce bacterial numbers, remove uterine debris and improve antibiotic penetration.

Breeding Management

The period of time after insemination and often times after ovulation is what we like to call the "critical" period. Often times owners are wanting to bring their mares home as quick as possible from the breeding facility after insemination or after ovulation is confirmed. The period of time immediately following insemination/ovulation is the window of opportunity to improve the mares chances of pregnancy. There are a few conditions that can happen during this time. We can assist the mare with these difficulties and often times make the difference in getting a pregnant mare or not. Some mares have difficulty evacuating the uterus post insemination or perhaps they mount an inflammatory process from the semen. These mares benefit greatly from assistance to evacuate the uterus and targeted treatment at reducing the inflammatory response.

Additional Diagnostics

Often times additional diagnostics are utilized to ensure proper treatment is instituted to improve a mare's fertility.

* Endoscopic examination of the uterus
* Endometrial biopsy
* Hormonal testing
* Endocrine testing

Treatments for Subfertile Mares

Laser cyst removal

Caslick procedure (surgical closing of the upper part of the vulva) to prevent contamination via aspiration of air, dirt, or debris

Uterine lavages

Uterine treatments

Ecbolics (medications to contract the uterus to improve clearance)

Anti-inflammatories

Antibiotics

Acupuncture

Hormonal treatments

Each mare is different and they are all treated as individuals. It is important to identity exactly what is impairing a particular mares fertility and target direct treatment towards resolving the problem(s).

MARE FERTILITY PROBLEMS CAUSED BY ENDOMETRITIS

Endometritis, an inflammation of the interior lining of the uterus, continues to be a major cause of subfertility or infertility in the mare.

There are many diagnostic tools that have been developed to determine if a mare has endometritis. These include ultrasonography, examination of the cervix with a speculum, examining vaginal discharge, uterine culture, cytology of the uterus, and endometrial biopsy. The most common methods are culture and cytology.

Endometritis is generally divided into infectious and non-infectious causes. Infectious endometritis is the result of bacterial invasion into the uterus. Treatment is directed toward eliminating bacteria with suitable antibiotics, and most infections respond well to this treatment.

Recurrent infectious endometritis has been suggested to be caused by bacterial biofilms or an invasion of bacteria into the tissue. Bacteria can survive in a dormant state under these conditions for a long time without being sensitive to antibiotic treatment.

Non-infectious endometritis is usually in response to a mare being bred or artificially inseminated. This is called post-breeding endometritis. When sperm are placed in the uterus, within a few hours a small portion of the ejaculated or inseminated sperm reach the oviduct. However, the majority of the sperm is eliminated and fails to reach the oviduct. Mares that remove all products of breeding from the uterus within 24 hours are considered as having a normal reaction to semen, whereas mares that do not remove the products are considered susceptible mares to persistent post-breeding endometritis.

The elimination of sperm after breeding is a combination of immune response and mechanical clearance through uterine contraction. Sperm in a mare's uterus trigger an immune response, which results in neutrophils (white blood cells) coming into the uterine lumen. These neutrophils bind with sperm and phagocytize (kill) the sperm. Also during the activation of the neutrophils, a hormone (prostaglandin F2 alpha) is released that causes the contraction of smooth muscles in the lining of the uterus. These uterine contractions assist in removing accumulated fluid and inflammatory products from the uterus.

In a normal, endometritis-resistant mare, once the products are removed from the uterus the inflammation subsides and the uterine environment returns to its normal state. In about 10-15% of mares, there is failure in the uterine defense mechanism to eliminate these inflammatory products and the mare is considered susceptible to endometritis.

Current Research

Elizabeth Woodward, a graduate student at the Gluck Equine Research Center, is examining the mechanism of impaired uterine clearance of susceptible mares after breeding. Previous studies have shown that impaired uterine clearance in susceptible mares was caused by reduced myometrial contraction in response to inflammation. However, the exact mechanism is not understood.

A recent study by the University of Kentucky Gluck Equine Research Center Director Mats Troedsson, DVM, PhD, Dipl. ACT, reported that susceptible mares had increased uterine accumulation of nitric oxide in the uterine lumen 13 hours after insemination. The nitric oxide causes smooth muscle relaxation and could possibly be the explanation for reduced uterine contraction in susceptible mares.

Woodward is conducting a follow-up study in which she is comparing nitric oxide levels in both resistant and susceptible mares after breeding. The goal is to identify 12 mares—six resistant and six susceptible—for use in this study. Each mare will then be challenged with insemination of dead sperm and evaluated at two, six, 12, and 24 hours after insemination. Measurements taken will include uterine fluid as measured by ultrasonography, number of neutrophils, and the level of nitric oxide in uterine secretions. Biopsies will be taken from the lining of the uterus and examined for nitric oxide synthase (iNOS) levels in the tissue.

Treatment options for sub fertile mares (Proceedings)

There are probably as many opinions as there are veterinarians in what constitutes an appropriate set of diagnostic procedures and therapeutic options for a subfertile or problem mare. This summary will describe the routine procedures that are performed in my practice and some of the most basic treatment protocols. It is emphasized that although some blanket therapies are established for certain routine conditions, the daily examination as well as the clinical signs will dictate changes or adjustments to the treatment protocol.

DIAGNOSIS

It is well established that a high percentage of mares that do not become pregnant have some degree of uterine contamination. It must be assumed that the uterine environment is a sterile environment that should be free of inflammatory cells. In a recent study by Riddle et al in 2006 it was reported that mares with bacterial uterine growth and/or cytological evidence of inflammation had a reduced pregnancy rate. In my practice, uterine culture and cytological evaluation of the uterus is the first diagnostic procedure that is performed. Case selection for performing a culture and cytological evaluation include: i) Mares that ovulate with a high degree of edema, ii) Mares that have free uterine fluid that is present throughout the estrus period, iii) Mares with short or long luteal phases, iv) Mares that have failed to conceive after being bred under good breeding management conditions with a stallion of good fertility v) Mares with evidence of uterine or vulvar discharge.

Mares that have a positive culture (isolation of micro-organisms) or a positive cytology (greater than 2 neutrophils/field) when appropriate techniques are used, should be treated. Appropriate techniques include but are not limited to taking the swabs when there is uterine edema and/or free uterine fluid, clean aseptic techniques and proper time of contact between swab and endometrial surface. When swabs are negative and an infection is suspected, a uterine biopsy specimen is submitted for culture or a low volume uterine flush is performed and a swab submitted from that specimen.

The cervix must be examined carefully to ensure that it is patent. Endoscopic evaluation of the uterus may be performed to determine if there are adhesions or foreign material.

Treating and breeding a problem mare

In order to establish an appropriate therapy in addition to an uterine culture and cytology, the degree of uterine edema and the presence or absence of uterine fluid is evaluated. If fluid is present the relative amount should be recorded, as should the degree of cervical relaxation and stage of the cycle. For example, a mare with heavy uterine edema and no follicles should be considered abnormal, while a mare with a large preovulatory follicle and no edema and a tight cervix could be in diestrus.

Treatment for bacterial uterine infections

There is no standard or blanket therapies in my practice and all mares are treated based on clinical signs. In order to minimize the interval from treatment to breeding it is critical that the previous ovulation date is accurately recorded in order to short cycle the mare with prostaglandin after treatment.

Mares with bacterial growth from uterine culture will be infused daily for 3 to 5 days with an appropriate antibiotic in a volume not to exceed 50 mls. A rectal ultrasonographic exam is performed daily in treated mares to evaluate the amount of uterine fluid accumulation. If fluid accumulation is detected either before the first intra-uterine infusion or during the treatment period a uterine lavage consisting of lactated ringers is performed. All efforts are made to recover the uterine lavage fluid. In an effort to maintain the uterus free of fluid, mares that are infused with intra-uterine antibiotics are administered carbetocin or oxytocin respectively two to four times daily. This is particularly important in those mare that fail to relax or open the cervix completely. Every effort is made to treat mares during the follicular phase of the cycle so that cervical relaxation, white blood cell influx and uterine contractility can assist in uterine evacuation. Antibiotics are given systemically to mares when the bacteria recovered is resistant to antibiotics that can be safely infused into the uterus, (i.e. enrofloxacin). Systemic antibiotics may also be used in mares with poor perineal conformation that require a constant vulvar seal maintained by the vulvoplasty (Caslick), and in mares with a pendulous, dependent uterus that does not drain fluid readily.

The degree of uterine edema is monitored daily by ultrasonography. Mares that ovulate with a high degree of edema or with edema that persists for more than 24 hrs post ovulation are considered abnormal. An endometrial biopsy sample is taken from mares with abnormal uterine edema to determine the degree of uterine inflammation if they are to be bred with expensive frozen semen.

Mares treated for endometritis are usually bred on the prostaglandin induced cycle after the treatment period without obtaining a second uterine culture. However these mares are treated examined 4-12 hrs after insemination and uterine lavage and ecbolic agents are used after breeding.

Mares considered to be problem breeders are only inseminated once preferably with good quality semen. Mares bred with fresh semen are mated in the 48 hrs before ovulation, within 24 of being bred with cooled semen and within the 4 hr interval after ovulation with frozen semen. Frozen semen is deposited at the tip of the uterine horn ipsilateral to the dominant follicle by rectally guided insemination. Cooled semen of mediocre quality is centrifuged and the mare inseminated with 2-2.5 mls of centrifuged semen also deposited by deep horn insemination. To insure ovulation, mares are given an ovulatory inducing agent. All mares are evaluated between 4 and 8 hrs post insemination to evaluate the amount of fluid accumulation and degree of edema. If free fluid is detected a uterine lavage is performed using warm ringers lactate until the efflux is clear. Uterine lavage is performed regardless of the ovulation status. Waiting for the mare to ovulate, may result in a prolonged uterine inflammatory process that could be difficult to control particularly if the mare is bred 24-48 hrs prior to ovulation. If no fluid is detected but the degree of uterine edema has increased the mare is treated with 20 mgs of Dexamethasone IM, and an ecbolic agent such as oxytocin, carbetocin is used for no more than 3 consecutive days. Mares are examined on a daily basis thereafter in order to establish day of ovulation and to evaluate uterine contents and edema pattern. A Caslick's procedure is performed on most problem mares since not all mares need to have a sloped vulva to aspirate air or become contaminated with fecal material. Many mares with a small ano-genital distance (anus to dorsal vulvar commissure) are easily contaminated.

4

Infertility and Subfertilty in the eve and DOE (Female Goat)

Luckily, infertility in bucks is rare. An inability to get does pregnant is usually related to nutritional deficiencies, which is why a good mineral is essential for bucks. As a buck gets older, his sperm count may go down, meaning he can service fewer does in a day, which is something to keep in mind if you have two does go into heat at the same time. Older bucks may also start to have problems with arthritis or other aches and pains that make it difficult for them to mount a doe.

If a young buck has been bred to multiple does and never settled any of them, a visit to a veterinarian is in order to make sure he is genetically a buck. An intersex goat may appear to be a buck on the outside, even though lacking all of the necessary anatomy.

GENETIC INABILITY

A lot of people worry about the ability of their does to get pregnant. The reality, however, is that less than 1 percent of does have a genetic inability to get pregnant, so it isn't something that you are likely to experience unless you have a sizable herd. Free-martins and hermaphrodites are very uncommon in goats, but a quick physical exam by your vet or anyone trained in artificial insemination will tell you if your doe is physically capable of getting pregnant.

The most common reason goats do not get pregnant is a mineral deficiency, and this is fairly common. Copper and selenium both play an important role in a goat's ability to come into heat, get pregnant, and stay pregnant for five months. If several goats in the herd are having fertility problems, you should look at your feeding and supplement program to ensure that it is providing adequate amounts of copper and selenium.

Ovarian cysts

It is possible for does to have cysts on their ovaries, which could keep them from settling, but it is not common. A cystic doe may appear to come into heat every week or not at all. This can be tricky for an inexperienced goat keeper, who may not notice a doe coming into heat if it isn't one of the more vocal ones. A cystic doe can be treated with hormone injections available from your veterinarian.

In does that have freshened before, it is possible that a sub-clinical uterine infection from the previous delivery is keeping her from getting pregnant or staying pregnant. This is more likely to occur following an assisted delivery where the attendant had their hand inside the uterus. Some people will automatically administer antibiotics to a doe after an assisted delivery with the assumption that it will take care of the risk of infection. A low-grade infection, however, may not have any other outward symptoms, so you would have no idea whether the antibiotics had worked until you found yourself with a doe that wasn't getting pregnant.

Goat Reproduction Reproductive Failure

Reproductive failure can be defined as either sterility or infertility. Sterility occurs when there is a permanent problem preventing kid production, while infertility, or temporary sterility, is the inability to produce live young within a specific time period. There are a number of factors that can cause reproductive failure including physiological, toxicological and psychological factors. Please see the links below for specific information on these causes.

In addition, there can be causes of reproduction problems that are inherited (passed on from parents). For example, females might have malformed reproductive tracts like incomplete oviducts, an abnormal uterus or a blind cervix or vagina. These problems may not detected without slaughter because the female may cycle (come in heat) regularly, but cannot get pregnant. Males may also have inherited abnormal problems as well.

More problems with reproductive anatomy might be the case of hermaphrodites. They are usually classified as true hermaphrodites or pseudohermaphrodytes. The true hermaphrodite will have both male and female reproductive gonads (ovaries and testis) while a pseudohermaphrodite will have either testis or ovaries, but have other parts representing both male and female.

Cryptorchidism is another condition that might lead to reproductive failure in males. This is a condition in which the testes do not pass through the inguinal canal into the scrotom, so they stay up in the body cavity. A male will be fertile in cases where one testis descends (unilateral cryptorchid), or even if part of one testicle is outside the body cavity, but if both testicles are still inside the body (bilateral cryptorchid), the male will be sterile. This condition is heritable (passed on from parents), so bucks with this condition should not be used for breeding.

Injury can also cause reproductive failure in both males and females. For instance, a female can be injured during kidding and a tear in the uterus may heal but may not allow the female to get or stay pregnant in the future. Injuries to the vagina, cervix or even vulva may also happen during kidding, such as if too much force is used to pull kids during a difficult birth. Injuries to the buck include inflammation of the scrotum/testes or injuries to the penis during natural mating or semen collection for artificial insemination (AI).

Goat Reproductive Failure Physiological Factors

Reproductive failure can result when hormones are not produced in the correct amount or in the right pattern or if no egg is produced and released from the ovary. Ovaries can become cystic (both follicular and luteal, so either before or after ovulation). In does with follicular cysts, which are rare, the follicle with the egg grows but there is failure of ovulation, or release of the egg. This type of cyst usually results in females that are always in heat. This may be related to problems with the hormone that causes ovulation known as luteinizing hormone. Even though this type of cyst is hard to treat, large doses of LH/hCG or GnRH followed by a dose of prostaglandin 10-14 days later, might be effective. See your veterinarian for use of drugs not labeled for goats.

Luteal cysts can occur when the corpus luteum (CL) fails to regress (go away) and becomes filled with fluid such that hormone secretion is altered. Poor uterine health resulting in low prostaglandin secretion could cause this problem. Luteal/CL cysts can lead to pregnancy loss, a dead, mummified fetus, infection, and even pseudopregnancy (fake pregnancy). Luteal cysts usually go away eventually, but prostaglandin (available commercially as Lutalyse©) administration might be used to lyse the CL.

Reproductive failure can also occur if external factors causing anestrus (no estrus/heat). These factors include season, low nutrition, lactation and disease. However, proper management or care can be used to overcome or lessen the impact of these types of reproductive failure.

It is also important to understand that females might undergo silent estrus, which is ovulation without visible signs of estrus. The incidence of a silent estrus occurring is high during transition periods in goats, such as the onset of puberty and transition into and out of the breeding season. In addition, reproductive failure can occur from females having shorter cycles, which has been associated with lower fertility. Females with longer cycles appear to be as fertile as females with normal cycle length.

Goat Reproductive Failure Toxicological Factors

When feeds get wet/damp, molds can grow and produce toxins, or mycotoxins, that can lead to reproductive problems and failure. Aspergillus, Fusarium and Penicillium are the most common toxins causing problems in stored feeds. In addition, there are molds and endophytes in forages that can also cause problems. Fescue toxicosis can results from endophyte (fungal)-infected tall fescue that make toxins such as ergot alkaloids, primarily ergovaline, that can delay onset of puberty, impair luteal function to result in reproductive failure, decrease milk production or fetal birth weights and cause longer than normal pregnancies. Cotton seeds, roots and leaves have a natural compound called gossypol that can cause a problem, mostly if producers feed whole cottonseed. Gossypolcan cause delayed puberty in young animals due to decreased growth. It can also cause problems with sperm production in males.

There are some noxious forages and weeds that can cause pregnancy loss or birth of abnormal kids. Lupins in the western United States and hemlocks, which are found everywhere, can cause defects such as cleft palates that can keep kids from being able to nurse. Poisoning from locoweed, found in the West, can cause bone defects and heart failure in the fetus. Many plants, such as soybeans, subterranean clovers and alfalfa, produce hormonal compounds called phytoestrogens that may reduce fertility in animals ingesting them. These compounds have been reported to cause weak or prolonged estrus, lack of estrus, irregular estrous cycle lengths, cystic ovaries and abortion.

Hormonal Causes of Infertility in the Doe

Infertility in does falls into many categories: hormonal, infectious, post parturition, chronic illness, malnutrition, old age and congenital problems.

With all the possible reasons why a doe may not conceive, it seems a miracle that any do. Fortunately, infertility problems tend to be exceptions rather than the norm. Healthy, sound does conceive easily and give birth to a high percentage of twins, triplets and sometimes quads. The number of kids is dependent on many factors, including genetic makeup, uterine overcrowding, mother's age, time of year the doe is bred, and so on.

Before we talk about infertility, we must learn something about the reproductive cycle. In most mammals, sexual receptivity is restricted to recurring periods called estrus. Does comes into heat, or estrus, every 18-24 days, often year round, depending on factors such as weather, nutrition and daylight hours. Pygmy does are precocious animals and can have their first estrus as early as six weeks old. Estrus typically lasts from 24-48 hours. Before and during estrus, hormones regulate changes in the does sex organs and make her receptive to a buck both physiologically and psychologically.

Signs of estrus are numerous, some obvious, some more discreet. The doe usually flags her tail side-to-side when around a buck, presumably to send attractive pheromones from the her reproductive tract into the environment that a buck finds attractive. Her vulva may be more pink than normal, appear swollen, and have some clear or white-colored discharge with the consistency of egg white. This discharge usually starts clear and becomes more white as the heat progresses. Others signs include more frequent urination and restless behavior. She may also talk more than usual, sometimes bleating very loudly at the edge of the fenceline nearest the buck. Decreased appetite and milk production are also reported. The doe is in a standing heat when she stands willingly and lets a buck mount. Standing heat usually lasts from 1-24 hours. If a buck is not present, does often mount their herdmates or stand for other does to mount them.

Ovulation is the most important event that takes place during estrus. This is a series of events, each triggering the next. A follicle on the ovary starts developing and producing estrogen. Increased estrogen triggers the release of follicle-stimulating hormone (FSH) and luteinizing hormone (LH) which matures the follicle. The number of follicles which develop during each estrus cycle depends on hereditary and environmental factors. The follicle, similar in appearance to a water blister, grows until it finally ruptures, releasing the egg. The egg, with its genetic contribution from the mother, is released from the ovary when the follicle bursts. The egg then floats down toward the oviduct. After ovulation, the collapsed follicle transforms into a new endocrine structure called the corpus luteum (CL) which produces progesterone to prepare the uterus for pregancy and to maintain the pregancy. If the egg meets healthy sperm in the oviduct/fallopian tube and fertilization takes place, the fertilized egg travels down the fallopian tube to the uterus where it implants itself.

When the doe conceives, the corpus luteum will remain throughout the pregnancy, secreting both estrogenic and progestational steroids. If the doe does not conceive, the uterus produces prostaglandin that dissolves the CL) leaving a small scar where the follicle was, and a new cycle begins.

The new cycle begins with a new follicle. Each doe is born with a certain number of follicles and when those follicles are used up, so is the possibility of pregnancy. This is called menopause in women.

To recap, a number of hormonal changes must take place for a doe to become pregnant. First, she must produce enough hormones to produce a follicle. The follicle must produce enough estrogen to trigger the release of LH and FSH, maturing the follicle and causing it to rupture. A corpus luteum must develop to prepare the uterus and maintain the pregnancy. Variance in any one of these areas can create problems.

Before you decide your doe has a conception problem, be sure she is not already pregnant. The most common cause of failure to cycle is pregnancy. Did a buck recently perform a great escape? Did a young buck stay with the herd too long? Did you record the breeding date(s) on a calendar? These questions need to be asked before giving any fertility treatments. Further, many times just exposing a doe to a buck (his odor) will bring her into heat. It is also possible for a bred doe to show signs of estrus, although her signs will not be as strong. This is where patience come in handy and maybe the key to treatment. But, if you have determined that your doe is definitely not pregnant, she was seen to be serviced and has still not settled, theres a list of hormone balance problems to consider.

Cystic ovaries occur when the follicles do not rupture and continue to produce estrogen. Heat cycles become become frequent (6-14 days) or nonexistent. Does may exhibit bucky behaviorby riding other does or being overly aggressive. If a doe with cystic ovaries is undiagnosed and untreated, she can become permanently sterile. These does eventually begin to look 'bucky.' If the condition is caught early, a shot of chorionic gonadotrophin may restore reproductive function.

If you suspect more than one doe of having cystic ovaries, your feeding program may be at fault. High estrogen levels in sweet clover and some alfalfa, or diets high in calcium and low in phosphorus may cause temporary problems.

There is evidence to support a hereditary predisposition to cystic ovaries from both the dam's side and the sire's side.

Another common theme is a 5-day cycle , a different condition than cystic ovaries. When this happens, the doe has a normal heat and the follicle matures but does not release the egg. The doe has another normal heat 5 to 7 days later, at which time the egg is released. There is diversity of opionion among experienced breeders about eggs not being released on the first of the 5-day cycles - some say that eggs can be released on both cycles and breed on both cycles to maximize the number of conceptuses. This condition may be related to the presence of multiple follicles - some rupture and some don't until the second heat.

If a doe has a hormone imbalance, she may exhibit no signs of heat at all, and the term for this is anestrus. The doe's ovaries are simply not producing follicles. A shot of prostaglandin may start her cycling. Anestrus may also be related to cystic ovaries. Silent heats between regular estrus cycles is not uncommon. This is where all the necessary physiologic and histologic events take place but the doe shows no outward signs of estrus. Fortunately, the buck always knows and a successful breeding may be accomplished if they are together. If the time between estrus periods is unusually long, silent heats may be the cause. [Ed. note: Anestrus for several-month periods may signal embryonic mortality if the doe had been bred.] The most probable cause is a lack of proper balance of hormones affecting the onset of estrus.

False pregnancy is another hormonal problem that can occur in pygmy goats. Pseudopregnancy is like normal pregnancy in every way. There are two ways to differentiate a normal pregnancy from a false one: (1) ultrasound or X-ray at around 60-90 days, or (2) wait until the end of pregnancy and see what is delivered. There is no known reason for pseudopregnancy but it may be a hormonal imbalance or early embryonic death with resorption of the fetus and membranes. If a false pregnancy is detected, a shot of prostaglandin will end it. Most does will not repeat this process and can be bred successfully on the next cycle.

Implantation failure in the uterus can create another infertile situation. If a fertilized egg fails to implant in the uterus, it simply deteriorates and a new cycle begins. Progesterone may play a big role in the preimplantation uterine changes. If progesterone is not released to prepare the uterus for pregnancy, the fertilized egg will simply be sloughed off. The uterine lining must be equipped for the egg to implant, like a lush shag carpet. The uterine lining of a doe that doesn't produce enough progesterone looks similar to an indoor-outdoor carpet. This type of uterine problem may be due to a uterine infection after the last delivery or an ageing uterus that finally fails. The prognosis is usually very guarded to negative for either of these conditions.

The occasional doe may have a retained corpus luteum that tells the doe she is still pregnant. This typically transpires after kidding and can be easily corrected with one or two injections of prostaglandin to restart normal cycles.

Detection of hormone problems is more difficult in Pygmy goats because of their small stature. With a cow or mare, a veterinarian can periodically feel the reproductive tract and follow its cycle. This is not possible with Pygmy goats. Many times trial and error is the best method. Your decision to pursue differing treatments should be based in part on the value of the animal. Is she worth the cost? Hundreds of dollars can be spent trying to identify the cause of an infertility problem. A series of hormonal screens can be done to either eliminate or identify potential problems, a laparoscopy can be done to explore the reproductive tract, or hormonal treatments can be given.

REPRODUCTIVE PROBLEMS OF DOES

Over-feeding, under-feeding, or feeding a nutritionally-imbalanced diet (especially one lacking in essential minerals such as copper, selenium, iodine, and zinc) will prevent a doe from conceiving. Poor nutrition also includes a diet lacking in energy, protein, and long fiber. Any or all of these conditions can prevent conception.

Overly-fat does may not breed, and if they do, may not be able to carry to term. Fat layers around internal organs (heart, liver, lungs, kidneys, and reproductive organs) and under the skin in goats; goats do not marble fat like cattle do. The Boer breed is especially prone to layering fat, so you need to watch processed feed input carefully. Overly-fat does have a much higher rate of pregnancy toxemia as they approach kidding. Once a goat is too fat, it is difficult to get fat off it, and putting a pregnant doe on a diet is fraught with danger. Don't let her get fat in the first place by learning how to feed properly.

A starving doe will abort her kids or deliver weak kids, if she breeds at all. The hormone leptin is well documented as preventing conception in improperly fed does, i.e. starving does, fat does, and does that are significantly underweight.

During drought in areas where pastures are normally plentiful, you must feed both hay and processed grains to your goats to offset these deficiencies. Feeding regimens must be adjusted from season to season, year to year, and even within a season as rainfall, temperature, and plant growth change.

If you are new to raising goats or if you are having problems with your existing herd, the first thing you should do is learn about proper nutrition. The most difficult part of raising goats in any managed herd is getting nutrition right. My website's Articles pages at www.tennesseemeatgoats.com contain valuable articles on mineral & vitamin deficiencies, forage/browse/pasture conditions, long fiber, processed grains, and plant/pasture toxicities.

Find a qualified goat nutritionist and enlist his services. The cost is minimal compared to the problems that you will avoid. The nutritionist does not have to be local, but he must know the nutritional needs of goats in your immediate area. Regional and national feed companies employ such people who should be available to you if you are a purchaser of their products. As a final alternative, contact someone whose knowledge you respect (but not a person who raises show goats unless you also raise show goats) and ask him for a referral.

Length of daylight impacts breeding cycles. A buck cannot breed a doe until she cycles into heat. Most breeds of goats cycle into heat only when the daylight begins to shorten. In the United States, this timeframe is from mid-July to mid-December. This is particularly true of dairy breeds but it also applies to some meat breeds under certain climatic conditions like extended daylight or darkness (Alaska, for example). Does usually come into estrus every 21 days during breeding season. If you wish to breed "off-season," you may have to use powerful estrus-inducing drugs in order to bring does into heat. If this is your choice, then veterinary supervision is recommended.

Occasionally a doe is born that cannot become pregnant. A freemartin is a sterile female sibling of a male. During fetal development, male hormones can be transferred through placental walls and render a female incapable of breeding. This is a relatively rare situation in most herds. Since I began raising goats in January 1990, I may have seen this one time in my herd.

Cystic ovaries can prevent conception. A doe that has not been put with a buck for several years after she has reached sexual maturity may not be able to breed if she has developed cystic ovaries. Breeding-age does need to be bred, but only once each year. Veterinary assistance is needed to diagnose this condition and determine if it is fixable. In many herds, the cost involved exceeds the value of the doe, so the goat is usually culled.

If your breeding program involves pasture breeding, you may incorrectly conclude that breeding is not happening if you don't see mating behavior taking place. Some goats in hot climates breed in the cooler night hours. Just because you don't see breeding activity does not mean that it isn't occurring.

A doe that is too old will not breed. Just as human females reach a timeframe beyond which they are fertile, so do female goats. When a doe reaches 9 to 10 years of age, she is less likely to become pregnant.

If an older doe does breed, extra care will be needed to make sure she carries successfully to term. She will have special nutritional needs and her udder may have so much wear and tear on it that her kids may need assistance in nursing. Older does tend to have stretched-out udders, teats that have become mis-shapen and difficult to nurse, less quantity and/or quality of milk, more chance of mastitis or congested udder, and a host of other issues common to older females. Commercial producers normally rotate out breeding does at five to six years of age and replace them with younger breeding-age females that are more likely to have less kidding issues and udder problems.

A doe that has been bred too young (less than 12 months of age) may experience dystocia (birthing difficulties), particularly if she has been unwisely bred to a buck of a larger-framed breed. If she survives kidding problems, she could have a torn uterus or other internal damage that can prevent future conception. You should wait until a doe is at least 10 to 12 months old before attempting to breed her. Always breed her to a buck of her breed or, if cross-breeding, use the services of a buck of a smaller-framed breed. Cattle producers long ago learned this valuable lesson after losing many calves out of cows bred to bigger-breed bulls to birthing problems. Many different abortion diseases can prevent conception or cause a doe to be unable to carry fetuses to term. Some of these diseases are zoonotic (transmissible to humans). Check the Articles page on my website www.tennesseemeatgoats.com for many articles of problems relating to pregnancy and kidding.

Injuries, illness, or heavy worm load can cause a doe not to breed. A doe needs to be in good physical condition and healthy to breed and give birth to healthy kids. Don't breed her back while she is nursing kids. Breed does only once per year.

Some does abort their fetuses naturally because of malformation during the pregnancy. Creating a life is a very complex process and lots of things can go wrong. Nature's way of handling these developmental failures is a spontaneous abortion.

Miscalculating a breeding date can lead to the conclusion that the doe has not bred. A sonnogram done at about 30 days of suspected pregnancy can confirm if a doe is bred and how many fetuses are present. However, for many producers, sonnogramming is not cost effective.A doe can experience a drug-induced abortion. The dewormer Valbazen and corticosteroids such as dexamethasone should not be used on pregnant does or does suspected of being bred. Think before using medications on does that may be or are pregnant.

Occasionally a doe can look pregnant (even appear pregnant on a sonnogram) and have a cloudburst or false pregnancy (hydrometra). She carries as if all is normal, even making a milk-filled udder, but has no fetuses and delivers only fluids. If this occurs in your herd, do not be too concerned unless it is a regular occurence, at which time veterinary assistance is required.

Many things can cause a doe not to breed and not to carry to term. You need to become familiar with all of these possibilities and be prepared to take both preventative and corrective actions.

MINERAL DEFICIENCIES AND INFERTILITY IN GOATS

It's not unusual this time of year to see people complaining on social media about infertility in their goats. Some people are quick to start injecting does with hormones, but many cases of infertility are caused by mineral deficiencies. If you are able to get a doe pregnant with hormones, she may still abort or miscarry or have other health problems if she's mineral deficient.

Copper

If you've been reading this blog for more than a month or two, you know I've written about copper a lot. Infertility is one of the more advanced symptoms of copper deficiency. In fact, it was one of the symptoms of copper deficiency that I should have noticed in our goats the second year we had them. One of the does would not get pregnant. The vet recommended Lutalyse for the doe, but she still did not get pregnant. The following year we had more goats that did not get pregnant.

If a doe doesn't get pregnant, she could be copper deficient if you are also seeing other symptoms of copper deficiency: faded coat, fish tail, balding around eyes or bridge of nose, parasite problems, or anemia that is not caused by parasite problems. If a doe gets pregnant, she may abort or give birth too early for the kids to survive. In severe cases, you may also see bowed legs, spinal injuries, and an overall weak immune system.

Copper deficiency is usually caused by minerals in well water, such as sulfur, iron, or calcium. If you are near a coal-fired power plant, that can leave sulfur deposits on your pasture. It can also be exacerbated by too much calcium or molybdenum in the diet, both of which are in alfalfa, which is part of the diet of most dairy goats. Alfalfa without well water is probably not a problem. But if you have well water and your goats get a high proportion of alfalfa in their diet, you will probably need to give more copper than you otherwise would.

If you have a doe that doesn't get pregnant, but she does not have a faded coat, fish tail, or balding on her face, it's probably not copper deficiency, especially if you don't have any of the issues that cause copper deficiency.

Selenium

Another cause of infertility in goats is selenium deficiency. This is why many people give BoSe injections prior to the start of breeding season. While copper deficiency is usually secondary, meaning that it's caused by copper antagonists, selenium deficiency is usually primary, meaning that goats are not getting enough selenium in their diet because most soil in the US is deficient in selenium. Secondary selenium deficiency can also occur, however, if you have a lot of calcium in your well water.

Goats that are selenium deficient may also have reduced fertility, meaning they are more likely to have single kids. If they get pregnant, they are more likely to have a retained placenta and get mastitis. It negatively affects their immune system, so they are also more likely to have problems with things like foot rot and parasites. If kids are born selenium deficient, they can't nurse. Selenium-deficient goats (and people) are more likely to have hypothyroidism, which in turn causes infertility.

Vitamin E

Symptoms of vitamin E deficiency are almost identical to symptoms of selenium deficiency, but the cause is entirely different. Goats usually become deficient in vitamin E when they don't have access to enough green food. This is usually only a problem towards the end of winter when the quality of the hay is deteriorating, or if you purchased hay that's not great quality. If you live on a small lot and don't have any pasture or browse, the quality of your hay is especially important.

Most people mistakenly think that BoSe is also a vitamin E supplement. It only contains a small amount of vitamin E as a preservative. The dose of vitamin E in it does not even meet a goat's requirement of the vitamin for a single day. MultiMin, which is another injectable selenium supplement, contains no vitamin E. So, if your goat has symptoms similar to selenium deficiency, but you've given the goat a dose of BoSe or MultiMin, you might suspect vitamin E deficiency.

MALE SCENT STIMULATES FEMALE GOATS' FERTILITY

The distinctive aroma of goats does more than just make barnyards extra fragrant. Male goats can use their heady scent to make female goats ovulate simply by being near them.

Researchers had ascribed this 'male effect' to chemicals known as primer pheromones — a chemical signal that can cause long-lasting physiological responses in the recipient. Examples of primer pheromones are rare in mammals; the male effect in goats and sheep, and a similar effect in mice and rats, where the presence of males can speed up puberty in females, are the only known cases. But exactly what substances are at work and how has remained a mystery.

Now, reproductive biologist Yukari Takeuchi from the University of Tokyo and her colleagues have identified a single molecule, known as 4-ethyloctanal, in the cocktail of male goat pheromones that activates the neural pathway that regulates reproduction in females[1]. "It has long been thought that pheromones have pivotal roles in reproductive success in mammals, but the mechanisms are scarcely known," says Takeuchi.

The researchers found that male goat pheromones are generally synthesized in the animal's head skin, so they designed a hat containing a material that captured their odorous molecules and placed them on the goats for a week to collect the scent. Analysis of the gases collected identified a range of compounds, many of which were unknown and were not present in castrated males. When exposed to a cocktail of 18 of these chemicals, the brains of female goats showed a sudden increase in the activity of the gonadotropin-releasing hormone (GnRH) pulse generator — the neural regulator of reproduction.

Citrus overtones

But one molecule stood out: 4-ethyloctanal, a chemical not previously found in nature and that has an orangy, floral odour. When presented to the female goats on its own, the chemical elicited a similar, albeit weaker, response, and the cocktail showed less of an effect when that ingredient was removed. None of the other chemicals appeared to have a statistically significant effect. The work is published today in Current Biology[1].

Peter Brennan, a physiologist at the University of Bristol, UK, says that the work will be useful in husbandry in goats and other ruminants, such as sheep, but he is not sure that the whole effect can be ascribed to the pheromones alone. "How much of this effect is innate and how much is learned?" he asks. Takeuchi admits that she cannot be sure, but she thinks it is an innate reaction, because it was seen irrespective of the mating experience of the female goats.

The main benefit of the work, says Takeuchi, is that it could be used to develop new, more-natural technologies that improve the efficiency of breeding and to treat reproductive disorders. "To control reproductive problems, it is important to regulate not only inhibitory factors such as infection or stress, but also accelerative factors such as pheromones," she says. The group is now looking to find similar pheromones and pathways in other economically important livestock animals, such as sheep and cows.

Goat Reproduction Puberty and Sexual Maturity

The male goat is called a "buck" or "billy." If he is castrated, he is called a "wether." Male goats up to 12 months of age are sometimes referred to as "bucklings." Adult male goats can weigh anywhere between 100 to 350 pounds, depending on their breed, health and nutritional status. Although they can come into puberty and breed does as early at 4 months of age, waiting until a buck is a year of age to start using him for breeding is best. The number of does a buck can breed during the breeding season is often referred to as "Buck Power" (Noble, 2004).

At 1 year of age, the buck should service no more than 10 does at a time (in one month). When he is 2 years old, he should be able to service 25 does at a time. At the age of 3 and older, he can breed up to 40 does at one time, as long as his health and nutritional needs are met. The number of does a buck can service at one time also depends on individual sex drive of the buck, the terrain of the land and if he is managed by a hand- or pasture- mating system. The buck has the greatest genetic impact on the herd and should be well taken care of at all times.

Day length has an effect on reproduction in the buck and the doe. Bucks have the highest libido (sex drive), fertility, and semen quality and volume in late summer and fall, the same time that of the doe (Senger 1984; Wildeus, date unknown). As the photoperiod, or day length, gets longer, less sperm is made and more abnormal sperm cells are found in the semen. During the fall, the endocrine system also increases levels of the sex hormones, testosterone and luteinizing hormone (Ritar,1990).

The Doe

The female goat is called a "doe" or "nanny." Up to the age of 12 months, she is sometimes referred to as a "doeling." The doe can reach puberty between 4 to 12 months of age, depending on the breed, season of birth, level of feeding/nutrition, and overall health status. Under-feeding can lower her chances of getting pregnant and having kids and can also reduce milk production after having kids.The genetic makeup of the animal also determines when puberty occurs in the female. Puberty is reached when the female exhibits her first heat (estrus) and ovulation.

Estrus, or heat, is the period in which the doe will stand and allow the buck to breed her. This phase of the reproductive cycle may last between 12 to 36 hours. The period from one heat cycle to the next is referred to as the estrous cycle. In goats, the estrous cycle occurs every 18 to 24 days, or 21 days on average. Does that are in heat may exhibit signs of mucous discharge from the vulva, swollen vulva, bleating, frequent tail wagging, pacing down the fence line and standing in heat. The doe can be bred when she has reached 60 percent to 70 percent of the average adult weight for her breed (80 pounds for an average Boer doe for example). Does that are bred too early may have problems during kidding (dystocia - kidding difficulty), and/ or their future reproductive performance may be harmed.

In season (when female goats are fertile): Between September and March; fertility lasts up to three days. If the female goat is not mated, she will come into season every 21 days. Once mated, goats can stay in milk for two to three years. Gestation (length of pregnancy) of a female goat : 150 days

Number of goat offspring: 1 to 3

Weaning (coming off milk onto solids): Goat kids are not usually weaned any earlier than 10 – 12 weeks. The young goats need to eat an increasing amount of solid food from 2 to 3 weeks old, in order to encourage rumen development.

Handling: Goats are sensitive, intelligent animals that should be handled in a smooth, calm manner.

Companionship: Goats are social and happier in pairs. They do have a 'pecking order'.

GOAT BREEDING

Does of any breed come into heat every 21 days for from 2 – 48 hours. A Doe in heat typically flags her tail often, stays near the buck if one is present. She becomes more vocal, and may also show a decrease in appetite and milk production for the duration of the heat.

Bucks also have a change in normal behaviour prior to breeding. They enter a period known as the 'Rut'. The 'Rut' is characterized by a decrease in appetite, obsessive interest in the Does, fighting between Bucks, a different display of behaviour and, most notably, a strong odour.

This odour is singular to Bucks in 'Rut', the Does do not have it unless the Buck has rubbed his scent onto them.

Gestation (length of pregnancy) is usually 150 days. At the end of the gestation period, either a single goat is born or twins or triplets. These are the most common births, although there are less frequent births of numbers higher than this.

Birthing is known as 'Kidding' and usually takes place with very few complications at all. After the female goat has given birth, she will eat the placenta which gives her most needed nutrients, helps to control blood flow and also reduces the birth scent to predators.

Following the birth, the Kid will sit somewhere safe while the mother goes to feed. When she returns, she will call to her Kid and it will come to her to play and be nursed.

After 'Kidding' the female goat will enter a period called 'Freshening'. This is where she begins her milk production. A healthy Doe will produce 6 pounds of milk per day while she is in 'milk'. The Doe will milk her Kid until it is ready for weaning at about 10 – 12 weeks old. Neutering early prevents the development of aggressive behaviour and nasty billy goat smells.

GOAT LIFE SPAN

Does (female goats) usually live on average 11 – 12 years. Premature endings for Does usually come through 'kidding' (giving birth). Does that are 'retired' from breeding around age 10 live longer, up to the age of 16 – 18 years. There was one incident where someone found a 24 year old female goat, which is quite a good age for this animal. However, it did state that she had retired from breeding at the age of 10, which goes to show, if you over breed your goat, it will not live too long, however, if you retire your goat from breeding at a reasonable age, you will have you goat for much longer. Bucks (male goats) actually live shorter than female goats. This is due to the fact that they go through what is known as a 'rut'. These periods of 'rut' is when they are ready to breed. Apparently it takes a lot out of them.

Some experiences in adapting previously free-ranging traditionally managed Matabele goats of Zimbabwe to individual stall-feeding

One hundred multiparous indigenous does were purchased from communal farmers for on-station trials. At the research site, the goats were confined and zero-grazed. During adaptation, six goats died, 28 aborted, 56 kidded end 10 had no reproduction records. The does lost weight and were afflicted by diseases not previously experienced under the traditional management system. After nine months of adaptation to the station regime, the goats were synchronised for oestrous and mated. Only 74% of the flock conceived and kidded while 19% did not conceive and 7% aborted. Factors considered to be associated with the poor adaptability to stall-feeding of previously freeranging goats are discussed.

Extensive goat production under the traditional communal grazing system is widely practiced in the semi-arid and arid regions of Africa which sustain 66.7% of the African goat population (FAO, 1984). This system, characterised by shortage of land, poor range management and seasonal fluctuation of feed supply favours mere survival rather than optimum production. Productivity of the goats in these areas can be improved by making more efficient use of the feed resources presently or potentially available to the farmer. To do this will require information on the nutrient quality of the feed as a function of season, the suitable physical form of the feed, the refusal rate and the physiological state and energy balance of the goats to be fed. Studies to generate such information often necessitate stallfeeding of the animals with no access to grazing. Unfortunately, when previously free-ranging goats are confined and stall-fed, they are stressed and may either eat less, lose weight, about (Wentzel, 1987) or even die (Kasowanjete et al, 1986). Many feeding trials are usually prematurely abandoned when goats fail to adapt to stall-feeding and unfortunately most of these futile efforts but nonetheless useful experiences are never reported.

Materials and methods

Location

The reported work was carried out at the former Thuli Breeding Station which is situated in Matabeleland Province in the south-western part of Zimbabwe. The region is semi-arid. The vegetation is mainly acacia shrub and mopane woodland with species of annual grasses appearing during the rainy months of November to March. Droughts are frequent, consequently herbage quality and quantity fluctuate within and between years. The total amount of rainfall during the study year (1989) was 162 mm. The mean maximum and minimum temperatures were 28 C and 13 C, respectively.

ANIMALS

In May 1989, one hundred multiparous indigenous Matabele goats were bought from communal farmers living within a 20-kilometre radius of the research station. The animals were driven on hoof to the research site. They were eartagged, weighed and their age was determined by counting the number of permanent incisors.

Management of the does during adaptation

On arrival the does were dewormed, dipped and vaccinated against pulpy kidney disease. Thereafter deworming and dipping were repeated on a regular basis.

The animals were housed in groups in a large animal house with concrete floors and partial roofing that allowed adequate ventilation. A forage mixture (two-thirds veld hay and one-third lucerne hay) with a mean crude protein content of 12% and neutral detergent fibre content of 60% was offered ad libitum in raised feeding troughs. Water and a salt mineral blocks were freely available to the animals. As from September 1989 (fourth month of adaptation), all the does were individually offered 100 9 maize grain daily so as to increase energy intake in preparation for breeding on-station. At the end of the fifth month of adaptation (October 1989), the goats were synchronized for oestrous using progesterone-impregnated intravaginal sheep sponges (Veramix 60 mg medroxyprogesterone, Upjohn, UK). The sponges were withdrawn after 14 days and 2 days thereafter, 10 large Matabele bucks, with a mean liveweight of 65 kg, brought in from Matopos Research Station, were introduced to the flock. The bucks continuously ran with the does during the day but were confined separately overnight so that mating records could be kept.

In order to determine pregnancy, blood was collected through the jugular vein 21 days after mating and twice weekly thereafter. The blood was kept at 2 C overnight and allowed to clot. After 24 hours, serum was decanted and stored at -20 C until assayed for progesterone levels using a radio-immunoassay technique developed for goat serum.

The adaptation period was intended to take three weeks but had to be extended to six months due to problems associated with feeding, disease incidences and reproduction.

Feeding

In the first four weeks of adaptation, feed refusals were approximately 60% of the feed offered as goats selected mainly lucerne hay which constituted only 24% of the total dry matter offered. When the forages were ground in an effort to reduce selection, feed refusal rates doubled because the feed was dusty. Low dry-matter intake (below 40% of the expected intake) during the first few months of adaptation is not uncommon in stall-feeding trials involving goats fed forage-based diets (Ademosun et al, 1985). The low feed intake has largely been attributed to the ability of goats to select and ingest the higher quality of diets on offer (Owen and Aboud, 1988). Owen and Aboud (1988) have shown that intake will improve when goats are permitted to reject at least 50% of the feed offered rather than the conventional 10 to 20%.

The reduced feed intake experienced in the first few weeks of adaptation may be the root cause of the observed liveweight losses, diseases and abortions.

Diseases

The most prevalent diseases during adaptation were off (contagious ecthyma), gaseous lymphadenitis, kerato-conjunctivitis (pink eye) and swollen knee joints. In cases of orf, daily swabbing of the lesions with salt solution and gentian violet led to rapid recovery.

In the second month of adaptation (July 1989) there was an outbreak of pink eye disease which spread rapidly amongst the whole herd; treatment with oxytetracycline led to quick recovery.

Swollen knee joints were observed in 20 does and all 100 breeding bucks in the second month of adaptation. These swellings were diagnosed as none pathogenic (no antibodies to caprine Arthritis encephalitis virus) and were attributed to lack of exercise due to confinement. The affected goats spontaneously recovered within four weeks.

Diseases encountered during adaptation may have led to the liveweight losses, abortions and acyclicity. In our study, incidences of orf and gaseous lymphadenitis were a lot higher than in normally experienced under the traditional system.

Six does died during the first month on station. Autopsies carried out were test negative for bacteriological and virological pathogens.

Doe liveweight changes

There were (P) differences in mean doe liveweights at buying between the different age groups (Table 1). Mature 8-tooth does were 18% and 35% heavier than the 6- and 4-tooth age groups, respectively. In the first 6 weeks on station all does lost an average 56 g/day. After six months adaptation (November 1989) when the does were mated, doe liveweight differences were smaller although the younger 4-tooth does were still significantly (P) lighter than the 6- and 8-tooth groups which were about equal in weight. As from November (sixth month of adaptation) all does gained weight. The highest gains were in the 4-tooth age group (90 g/day) compared to 53 g/d and 80 g/day for the 6- and 8-tooth groups. These weight gains could be attributed to replenishment of lost body tissue and genuine growth in the immature 4- and 6- teeth groups.

Reproduction

Not withstanding the concerted efforts to avoid buying pregnant females, about 90% of the goats purchased turned out to be pregnant (Figure 1), mostly in the second month of gestation. None of these goats had shown overt pregnancy symptoms when purchased.

During the first three months of adaptation, 30% of the does aborted. The highest number of aborters was in the young 4-tooth age group where almost 1.5 and 8 times more does aborted than in the 6- and 8-tooth age groups, respectively (Figure 1). Specimens from the aborted foetuses and placentae were tested for leptospirosis, contagious abortion, Rift Valley fever and blue tongue. All test results were negative and it was postulated that stress factors (such as a sudden change of environment, rather than disease could have caused the abortions. Based on information available in other ruminants and observations made in goat flocks, Morand-Fehr (1987) has concluded that pregnant goats are very sensitive to various kinds of stress such as rapid variations of environmental conditions, excessive animal density in the goat house, struggles between animals, psychic shock etc.

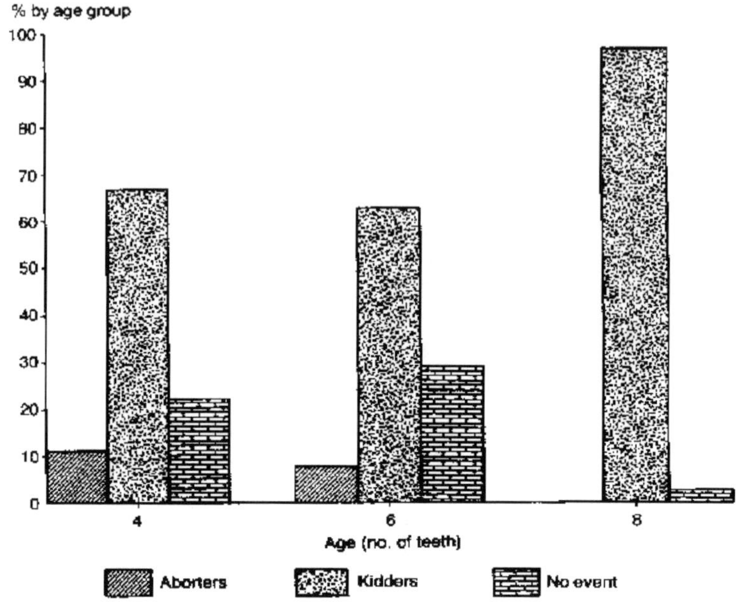

Figure 1. Reproductive response to breeding on station (April-July 1990).

Hypersensitivity to stress of pregnant goats has been explained by the theory that, unlike other domestic ruminants where both the corpus luteum and the placenta secrete progesterone, pregnancy in goats is sustained by progesterone secretion from the corpus luteum only (Sheldrick et al, 1980).

Intravaginal sponges that were used for oestrus synchronization could not be withdrawn from 10 does. The sponges were eventually pulled out after surgical manipulation of the vagina. Five of the does resumed normal oestrus after three months but the others remained acyclic.

Mating was expected to commence two days after sponge withdrawal but was delayed by 20 days. Reasons for this delay cannot be readily explained. It is possible that the sheep intravaginal sponges used were unsuitable for goats, though other workers have reported successes with progesterone impregnated sheep sponges used on goats (Cameron et al, 1988). Environmental effects such as ambient temperatures could have contributed to the delay of oestrus onset or lack of libido after sponge withdrawal. It has often been stated that goats in the tropics will kid all year round (Wilson et al, 1984; Hale, 1986). However, there are indications of definite peak kidding periods, usually two per year, availability as influenced by rainfall patterns (Hale, 1986). It is possible that the rains also have a cooling effect on cyclic goats which enhances libido. In sheep, it has been shown that a high proportion of mating occurs in the early morning and activity subsides to a low point at midday and increases again in the evening (Blockey and Cummings, 1970). In this study the mean daily temperature was above average (34 C) when the goats were exposed to the bucks and mating only commenced a day after the first rains of the season.

Mating (November 1989) lasted about four weeks and only 50 % (48 goats) of the flock conceived as confirmed by consistently high blood progesterone levels

(1 ng/ml). The rest of the flock which did not conceive in November continued to run with the bucks. Twenty eight of the remaining 46 goats were mated in February after which there were no further matings.

Combining the reproductive outcome for matings in November and February (after 6-9 months adaptation to on-station regime), 74% of the original flock purchased managed to conceive and kid, 7% aborted while 19% remained acyclic recording basal blood progesterone levels (<0.3 ng/ml). The number of goats that aborted was far less after nine months (7%) on-station, as compared to 28% in the second month on-station. Mature 8-teeth does had the best performance with the highest number of goats kidding (96%), the least number of does remaining empty (4%) and no aborters. The abortion rate was higher in the younger 4-tooth does (11%) than in the 6-tooth does (8%) which were 7.6 and 4.2 kg, respectively, does lighter than the mature 8-tooth does at mating. The presence of habitual aborters (does with a long standing poor reproductive history) in the flock cannot be ruled out. It is quite likely some farmers took advantage of the sale to rid themselves of their nonproductive stock. This could also explain barrenness of 10 goats at buying and the failure of the same goats to conceive at mating in November and in February.

The fact that more does conceived when mated in February (increasing the total kidding rate in response to breeding on-station from 46% to 74%) means that given time to adjust a larger proportion of the goats will conceive and kid. Unfortunately this prolonged adaptation greatly inflates the cost of goat research projects as compared to cattle and sheep which usually adapt to stall-feeding in 3 weeks. Unlike cattle and sheep, goats seem to have a lower stress threshold demonstrated by being very sensitive and reacting adversely in an exaggerated manner to stress caused by change of routine.

Conclusion

In our experience, adaptation to on-station regime by indigenous Matabele goats previously managed under the traditional extensive system took more than 6 months. In the first few months, the goats ate very little forage and consequently lost weight. Diseases not previously experienced in the traditional system became a problem on station. Reproductive performance was poor, especially among the younger groups of females, suggesting that adult females have a better chance of adapting to complete confinement.

It seems that numerous environmental factors acting either individually or collectively will impose some degree of stress but our knowledge of the ability of these previously freeranging goats to adapt or adjust to stall-feeding needs to be improved by further long-term research.

The Ultimate Guide to Breeding Goats

Most dairy goats are seasonal breeders.

Most "Alpine" breeds of goat will only breed during their breeding season. This is usually between the months of August and December. These breeds include most of your large dairy goats such as LaMancha, Saanen, Alpine, Oberhasli and Nubians.

Sometimes Nubians can be forced into year-round breeding, but that is unpredictable. Your miniature breeds such as Pygmy and Nigerian Dwarf, as well as meat breeds, like Boer, will breed year round. So when you are planning for your kids and milking, keep in mind this time restraint.

A doe should be at least 80 lbs before breeding

When it comes to breeding, it's not so much age as it is size. Most of your standard sized dairy goats will need to make the weight of about 80 lbs before they are bred. A healthy, well-fed doe should make this weight by about 8 months. Some people like to wait until they are about 1 1/2 years old before breeding but it isn't necessary.

I have seen no reduction in growth between breeding 8 month old does and 1 1/2 year old goats. Nubians particularly are known for being a little on the slower size when it comes to growth, so make sure you have a reliable way to weigh your goats. We don't have a livestock scale and for years we've made do with our bathroom scale. My husband just weighed himself and then picks up the goat and weighs himself again.A goat weigh tape is another good option that is very accurate.

Start keeping an eye on weight in August so that you can give additional nutritional support to any does that need it so that they can make weight before the season ends.

A doe's heat cycle is every 18-21 days

Starting in August your doe should start coming into heat every 18-21 days. Depending on the doe, she will stay in heat anywhere from a few hours to a couple days. Signs to look for:

- Tail flagging
- Clumped/Wet hair on the side of her tail
- Mucous discharge
- Swollen rear end
- Yelling (more than usual)
- Most obviously- interest in your buck or a "buck rag"

Dating vs Living in

When it comes to the actual breeding process you have a couple choices. You can take your doe and buck on "dates" or you can house them in the same pasture for a set period. There are pros and cons to both sides, but some things to consider are:

- Dates will give you much more control over the due dates and assurance that the deed did get done
- Sharing a pasture will reduce the risk that you miss the breeding window by not seeing the signs of heat soon enough
- Some does won't stand for a buck without help. In this instance, taking her on a lead to the buck will force the breeding to occur
- I've had a doe that showed ZERO signs of heat- unless the buck was present. By pasturing them together I could insure that she was bred.

The Gestation cycle of a goats is 150 days, or about 5 months.

When you are planning your breeding keep in mind that the gestation period is about 150 days for your standard sized breeds. If you live in a cold climate with harsh winters you probably will not want your kids being born in a snowy and cold January- which means don't breed your doe in August!

I like my kids all born sort of close so that they grow up together and in case of a single birth the lonely kid will have other kids to play and sleep with, so I don't stagger my breedings very much.

When you notice your doe in heat, write it down on the calender, then calculate the possible due date and jot that down as well (Here is the link to a goat due date calculator). As long as she doesn't come into heat again, you can use this approximate due date to plan her prenatal care.

Bucks stink...and other buckish behavior

If you have just one or 2 goats you may opt not to keep a buck on hand. But if you plan on breeding many does or there is not a suitable stud to rent in the area you will most likely end up with a buck on the premises.

And here's the thing about bucks- or at least a buck in rut- They stink. Let me say that again: Bucks STINK! Seriously. Young bucks aren't so bad, but the smell ripens with age.

Once breeding season hits and your does go into heat and your buck goes into rut they will start some rather amusing, and unsavory, behavior. Such as:

- Urinating. I know, that doesn't sound so bad. But they pee on their faces, on the back of their legs, in their mouth....It's rather amazing the reach they can achieve. Eventually their faces and legs will be covered. Good thing that the does find this irresistible.
- Blubbering, tongue flapping and lip raising. All these behaviors happen towards the does. Very amusing to watch his mating rituals!
- Aggression. Bucks will be more aggressive when they are in rut. If you have more than one buck take special care when there are does around so that they don't end up fighting. Also never turn your back on a buck in rut- even one you may have raised from a bottle. Even if they are not meaning to they can potentially hurt you- especially if they decide to "practice" on you

I actually find our bucks some of the sweetest of our goats. But when they are in rut, I feed them last and have a special coat I wear over my clothes to help minimize the smell contamination.

Keep them in Good Health

During breeding seasons your goats will need extra nutrition to support the stresses on their bodies. Heat, rut and pregnancy takes its toll on their bodies, so be sure to support them with extra grain, high quality hay and lots of forage.

Your bucks in rut and your does will need about 1 lb of grain a day. I give hay depending on the amount of forage- if it is rich and plentiful, I do not give any hay. If they are on a dry lot or small pasture they will need more supplemental hay. I also add black oil sunflower seeds to all of our goat's feed.

In addition to their feeds, breeding season is also a good time to make sure they are not deficient in any vitamins and minerals. We keep loose minerals out free choice for our goats, but if you live in an area that has deficient soil you may also have to give the following supplements:

- Copper– prevents kids being born with copper deficiency and gives your goats a stronger appetite and a healthier growth rate
- Selenium/Vitamin E– this prevents white muscle disease in kids and can increase fertility
- Nutridrench– All purpose vitamin for any goat that may need a little extra nutrition

Line Breeding vs. Inbreeding

When it comes to breeding, you want to make sure your keep a variety of genetics in your herd. There's a saying something along the lines of: It's called line-breeding if if works, in-breeding if it doesn't. And that pretty much sums up the topic in a very simple way.

The buck is actually a very important part of the breeding process. He is the easiest way to add in good genetic traits to your herd. When it comes to line-breeding there is no set rules such as breeding daughter and grandfather, except never breed full brothers and sisters. Occasionally you can breed father/daughter but it's not ideal.

Line-breeding will accentuate the good qualities- and the bad. If your buck has any negative traits (aggression, mother had low milk production,etc) those traits will be accentuated in his offspring if the are a product of line-breeding. The best way to get around this is to only keep or breed high quality bucks and if you keep your own bucks keep at least 2 on hand for the purpose of genetic diversity.

5

Reproduction in The Buffalo

Buffaloes are integral to Indian livestock sector followed by others ruminant species. India possesses 199.1 million cattle and 105.3 million buffaloes and ranks first in milk production in global world mainly contributed by buffaloes (DAHD, 2010). But the milk productivity is much lower due to various reproductive and non-reproductive disorders. The success of buffalo husbandry lies in optimal reproductive rhythm as well as disease control of animals round the year. Reproductive function may be affected by the management, disease and genetic constituent of the animal. Although much advancement have been made in controlling reproductive diseases in cattle and buffaloes, serious losses are still going on and as such infertility remains a major economic problem and its incidence appears to be rising in India also. However the impact of infertility in terms of economic losses in Indian livestock is not available.

Reproductive efficiency in buffaloes can be obtained by optimum fertility of buffaloes. A sound reproductive cyclicity is essential for ensuring regularity of calving paving way for profitable dairy farming. But this is hindered by the occurrence of various reproductive disorders from infectious and non-infectious sources. Reproductive disorders are multi-factorial problems which depending on their nature and severity produces varying degrees of suboptimal reproductive performance. Moreover these, there are various sources of non-reproductive or general health problems which have an adverse on the reproductive performance.

- The major reproductive problems affecting buffaloes are
- Anestrus
- Repeat breeding and
- Post parturient disorders.

The pre-parturient complications conditions lead to parturition associated problems as well as delayed involution of the uterus and metritis. Parturition associated problems include dystocia, still birth, abortions, retained placenta etc. Stillbirth and dystocia have been recognized as the most important factors compromising the future reproductive life of the animal. They increase the odds of developing metritis and retained placenta. Such conditions often cause delayed uterine involution and predispose to secondary infections and abnormalities in the resumption of ovarian cyclicity. For optimization of reproductive performance, the reproductive health and management in buffalo demands critical evaluation and monitoring of the present scenario in buffaloes and its therapeutic measures.

RECENT DEVELOPMENTS IN BUFFALO REPRODUCTION

The buffalo is an important livestock resource in several countries of South Asia and the Mediterranean regions. However, reproductive efficiency is compromised due to known problems of biological and management origins, such as lack of animal selection and poor nutrition. Under optimal conditions puberty is attained at 15 to 18 months in river buffalo, 21 to 24 months in swamp buffalo and is influenced by genotype, nutrition, management and climate. However, under field conditions these values deteriorate up to a significant extant. To improve reproductive efficiency, several protocols of oestrus and ovulation synchronization have been adopted from their use in commercial cattle production. These protocols yield encouraging pregnancy rates of (30% to 50%), which are comparable to those achieved in buffaloes bred at natural oestrus. The use of sexed semen in buffalo heifers also showed promising pregnancy rates (50%) when compared with conventional non-sexed semen. Assisted reproductive technologies have been transferred and adapted to buffalo but the efficiency of these technologies are low. However, these latest technologies offer the opportunity to accelerate the genetic gain in the buffalo industry after improving the technology and reducing its cost. Most buffaloes are kept under the small holder farming system in developing countries. Hence, future research should focus on simple, adoptable and impact- oriented approaches which identify the factors determining low fertility and oestrus behaviour in this species. Furthermore, role of kisspeptin needs to be explored in buffalo.

Buffalo has a significant role in the agricultural economy of many developing countries by providing milk, meat and draught power. The world population of buffalo is estimated to be 199 million (FAOSTAT, 2012) with more than 96% of the population located in Asia including 16.4% of Pakistan's contribution. In recent decades, buffalo farming has expanded widely in the Mediterranean and Latin America as well as, in Central/Northern Europe where several herds were introduced.

Dairy buffaloes have been used for milk production in India, Pakistan, some other South Asian countries, the Middle East and Italy; while dairy characteristics are being induced in the local population of Indo-Chinese Region and South America through crossbreeding with Pakistani Nili Ravi and Indian Murrah buffaloes. The milk yield increased from 700 to 2,000 kg/year in China through crossbreeding.

The buffalos can utilize poorer quality roughages, adapt to harsher environments and are more resistant to several bovine tropical diseases. Despite these merits, buffalo have relatively poor reproductive efficiency irrespective of their location throughout the world. Buffalo exhibit many of the known reproductive disorders including delayed onset of puberty, poor oestrus expression, longer postpartum ovarian quiescence, and most importantly lowered conception rates particularly when bred artificially. However, higher fertility could be achieved through better feeding and management. It appears that because buffalo are located mostly in developing countries with meager resources, there is limited quality research in the area of basic physiology, health, management, nutrition and applied reproduction.

The objective of this review is to examine the major recent developments in buffalo reproduction. We discuss the impact of the various techniques as well as bottlenecks and possible future developments which will lead to improve reproductive performance in this species.

PUBERTY

Buffalo usually attain puberty when they reach about 60% of their adult body weight (250 to 400) kg, but the age at which they attain puberty can be highly variable, ranging from 18 to 46 months (Jainudeen and Hafez, 1993). The factors that influence this are genotype, nutrition, management and climate. It could be attained under optimized conditions at 15 to 18 months in river buffalo and 21 to 24 months in swamp buffalo (Borghese, 2005). The delay in puberty, consequently delays conception and results in low reproductive efficiency and lengthening of the non-productive life. A major cause of delayed puberty may be poor feeding and management under field conditions.

OESTROUS CYCLE

In order to enhance reproductive efficiency of buffalo, a thorough understanding of the regulatory mechanisms involved in the oestrus cycle is required. The duration of the oestrous cycle in buffalo is similar to that in cattle, ranging from 17 to 26 days with a mean of around 21 days.. However, there is a greater variability of the oestrous cycle length in buffalo compared to cattle, with a greater incidence of both abnormally short and long oestrous cycles. This may be attributed to various factors including adverse environmental conditions, nutrition and irregularities in secretion of ovarian steroid hormones.

In buffaloes, ovarian follicular dynamics during the oestrous cycle is similar to that in cattle. Studies from India,Brazil and Pakistan have shown clearly that the majority of buffalo have two waves of follicular activity during their oestrous cycle. More investigations on the effect of follicle stimulating hormone and nutrition on number of follicular waves need to be studied in buffaloes.

Studies on oestrous behavior and endocrinology in buffalo (Roy and Prakash, 2009; Singh et al., 2000) indicate considerable variations in reproductive endocrine activity without external signs of oestrus (silent heat) are common. The low intensity of oestrus in buffaloes may be due to low circulating concentrations of 17-â oestradiol in comparison with dairy cattle (Seren et al., 1995).

Furthermore, tying up the animals as per normal husbandry practices in many developing countries restricts the ability of buffalo farmers to observe heat signs (Warriach et al., 2009). Buffalos also tend to show heat signs during the night when farmers are not observing their animals (Unpublished data). Season is another extrinsic factors that influences the characteristics of oestrous behaviour. In the tropics, high ambient temperature reduces sexual activity during the day (Jainudeen, 1977) and shortens the oestrous period (Gill et al., 1973) with the incidence of silent oestrous more common during the hot summer season. These adverse effects of heat stress make oestrous detection much more difficult in buffalo. Oestrous detection could be significantly improved through the introduction of a teaser bull or an androgenzied female (Chohan et al., 1992).

The interval between standing oestrous and ovulation, which is very important for artificial insemination, was 30 hours in buffaloes (Warriach et al., 2008). Under field conditions, the am-pm rule of insemination originally developed for cattle (Trimberger, 1948) is generally followed in buffaloes. To follow this rule, the buffaloes should be bred 12 h after the detection of standing oestrus. However, onset of heat signs instead of onset of standing oestrus has been erroneously considered as the land mark with buffaloes often being inseminated, earlier than required. This early breeding is potentially responsible for lowered fertility, and can be explained by the fact, there is an interval of about 8 to 10 h between onset of heat signs and onset of standing oestrus. This indicates buffaloes should be inseminated 12 h after the detection of standing oestrus (detection by bull/teaser) or alternatively 18 to 24 h after the onset of heat signs. In order to confirm this approach, investigations are required on the timing of insemination in relation to standing oestrus and pregnancy rate.

SEXED FROZEN SEMEN

Semen sexing has been successfully used for producing living offspring in bovine species (Seidel et al., 1999). In buffalo, a difference in DNA content between X and Y sperm was found, and based on this difference it has been further demonstrated that processing buffalo semen was feasible (Lu et al., 2007). In a recent study, promising pregnancy rates (50%) were achieved when inseminating a dose of sexed semen containing 4 million spermatozoa (Gaviraghi et al., 2013). In order to expand use of this technology, there is a need to further refine this protocol for buffalo breeds of commercial significance such as the Nili-Ravi buffalo of Pakistan.

ASSISTED REPRODUCTIVE TECHNOLOGIES

The first successful embryo transfer in buffalo was performed in the United States of America (Drost et al., 1983). Subsequent successful transfers have been reported from many other countries. However, the success rate is much lower in buffaloes, due to their inherently low fertility and poor superovulatory response (Misra et al., 1990). The average yield of transferable embryos is less than one per superovulated donor. The buffalo ovary has a smaller population of recruitable follicles at any given time; an average of 12,000 primary follicles has been reported (Danell, 1987), compared to the average in the ovary of the cow, which has an average of 133,000 (Erickson, 1966). This technique comprises a series of carefully integrated sequential steps including donor selection, donor treatment, recipient selection, insemination of the donor, embryo recovery, embryo handling and evaluation, embryo transfer, and recipient care. The technology has to be refined to account for the lower, less responsive follicle population in the buffalo.

Assisted reproductive technologies have been introduced to overcome the inherent reproductive problems, fast propagation of superior germplasm, and to reduce the generation intervals. These technologies provide an excellent source of embryos for carrying on basic research in developmental physiology, farm animal breeding, and for commercial application of the emerging bio techniques like cloning and transgenesis. During the past two decades, considerable advances have been made in our understanding of buffalo reproductive physiology, however, previous reviews (Palta and Chauhan, 1998; Gasparrini, 2002) and more recent studies suggest that the rate of transferable embryo yield remains at a plateau (Manjunatha et al., 2009). Results have been quite variable between laboratories and are most likely related to differences in embryo physiology, metabolism, and culture requirements among buffalo breeds. Further studies are also needed to improve the cryopreservation of *in vitro* embryo production embryos.

CONCLUSIONS

Buffaloes are an important livestock resource for many countries. Most buffaloes are kept under the small holder farming system in developing countries. Future research should focus on simple, adoptable and impact oriented approaches which identify the factors limiting fertility and oestrus behaviour in this commercially significant species.

Despite the inherited problems in buffalo slow progress has been made in the application of assisted reproductive techniques. Artificial insemination is practiced commercially; embryo transfer, in vitro embryo production, and nucleus transfer remain in the realm of experimentation. If their costs are reduced these latest techniques offer the opportunity to accelerate the genetic gain in the buffalo industry with the proviso that they are used in conjunction with efficient national progeny testing and sire evaluation programs.

BUFFALO REPRODUCTION IN INDIA

The buffalo is backbone of the farmer's economy of India, South-Asia and several European and American countries, benefiting nearly half of humanity in over 40 countries. The Asian countries are producing over 96 % of 48 million tons of the world's total buffalo milk output at an annual growth rate of 4 % (Sasaki, 1997). Buffaloes contribute more than one-third of total milk production in Asia and are the second largest producer of milk in the world. Buffaloes are also valued for meat and draught purposes (Bandyopadhyay *et al.*, 2003).

The world buffalo population is estimated to be 166.4 millions spread in some 129 countries around the world (FAO 2002); 161.4 million buffaloes are found in Asia (97.2%), 3.55 millions in Africa (2.13%), 1.4 millions in South America, and 0.257 million in Europe (0.154%). India has over 93.13million, or approximately 56.6% of the total world buffalo population (FAO, 2001).

Since 1981-82, the number of buffaloes has increased by a total of 38.8 million which is a 30.4% increase. Interest in water-buffalo breeding is growing in several countries where here to fore the buffalo has been neglected and even unknown, especially in South-America and in Europe.

River-buffaloes are concentrated mainly in South-Asian countries like India and Pakistan. They are reared mainly for milk production with meat as an adjunct. Buffalo bulls are important source of draught power. Buffalo meat is of good dietary value; it is lean and contains less saturated fat compared to beef and pork. In India, buffalo contributes 34.7% of total meat production (Murphy and Prince Davadaso2003). Presently, the buffalo contributes 1.41 million metric tons of meat to the meat pool of India, which comes out to be 45.74% of global buffalo meat. (FAO, 2000).

The buffalo is a difficult breeder because of its inherent susceptibility to environmental stress, which causes anoetrus and sub-estrus. These two conditions are responsible for a prolonged intercalving period resulting in great economic losses for the dairy industry. The productive and reproductive efficiency of animals are complementary to each other. Low reproductive efficiency in general and in the buffalo in particular remains a major economic problem globally and its incidence is higher in our country. The main interest of the animal breeder is to achieve more young ones in a lifetime, reduced mortality, and healthy and superior young ones. To achieve this goal, normal reproductive tools to augment reproductive efficiency (fertility), means to overcome reproductive constraints, and remedies are needed. The present article reviews reproduction in buffaloes and different approaches to augment reproduction.

Normal Reproduction in Buffaloes:

The age of puberty in buffalo is 36 to 42 months in India. It is somewhat late as compared to other countries like Italy, where the age at first calving is 36 months on average (Borghese *et al.*, 2003). The estrus cycle length is 21 days with a heat duration of 12-24 h. The ideal buffalo produces a calf every 13 to 14 months.

The factor which most strongly influences age at puberty is nutrition level. Buffalo exhibit seasonablity in reproductive activity. The reproductive cycle of buffalo is as follows (Bhikane and Kawitkar, 2000)

- Age of puberty: 36-42 months
- Length of estrous cycle: 21 days.
- Duration of heat: 12-24 h.
- Time of ovulation: 10-14 h after end of estrous.
- Period of maximum fertility: Last 8 h of estrous.
- Gestation period: 310 days.
- Period of involution of uterus: 25-35 days.
- Breeding season: September to February.

Breeding systems in buffaloes:

Natural mating: Generally in dairy buffaloes, natural mating is used. In certain states, better awareness the buffalo bull and of the dam's milk yield are used in breeding. But for the most part there is no consideration of pedigree and any potent entire male available at estrous is used for breeding. This practice has cause a great deterioration in performance of the buffalo and the destruction of breed characteristics.

Artificial Insemination: Both fresh and short- duration stored extended liquid semen as well as frozen semen is being used for insemination of buffaloes. This service is provided on nominal paymentby state department of veterinary medicine and animal husbandry and other agencies like BAIF and NDDB. The network project on buffalo development of ICAR is arranges breeding with the use of frozen semen of known pedigree; Murrah, Nili-Ravi, Surti, Bhadwari, Pandharpuri and the swamp buffaloes of Assam are included in this programme at present (Pathak, 2003).

Although the density of good breeds of dairy type river-buffaloes is greater in the hot and hot-humid parts of tropical India, the problem of low fertility in summer is also more serious. Some of the main reproductive problems of economic importance are:

1. Delayed maturity: Delayed maturity in both male and female buffaloes is common throughout India. This is due to neglect of calves during there growing period. Buffaloes have potential to gain at rates of 400-800 gm daily after about 4-6 months of age and can attain 300-450 kg body weight suitable for breeding at about 24 months of age. But in the majority of dairy buffaloes calving occurs at 4-6 years of age. This is due to an inadequate supply of feed and nutrients during the growing phase (Ranjhan and Pathak, 1992).

2. High incidence of silent estrous and short duration estrous: This is more serious problemduring the hot and humid-hot months. Due to these reasons, heat is often missed, and tethered females or females grazing with outpotent males in the herd remain unbred. This is an important factors contributing to the long intercalving period.

3. Non-availability of proven sires and good quality semen of proven buffalo-bulls.

4. Occurrence of venereal diseases causing infertility and sterility: These problems are encountered in both sexes, and the cause of there spread is the use of bulls for natural service. In recent years these problems have shown an increasing pattern because of increased use of artificial insemination in remote villages with the help of improperly trained inseminators. In some places, such persons present themselves as veterinary doctors though they have no education clinical training. These persons may cause severe damage to livestock health and the wealth of the country. (Pathak, 2003)

5. Poor heat sign: Silent heat in the buffalo is one of the most important unsolved impediments to efficient breeding. It occurs in the hot seasons. A combination of estrus detection methods may be necessary for identification of animals in heat. (Remesh *et al.*, 2002).

6. Anoestrum: Post-partum anoestrus in some buffaloes that have calved during cool season, i.e. the normal breeding season, have a shorter post-partum estrus interval than those that calve during last breeding season (February through July).

7. Poor nutrition: Adequate nutrition is prerequisite for proper functioning of the reproductive system in animals. Underfeeding, overfeeding; protein and vitamin deficiency, and imbalance of trace elements results in various reproductive abnormalities.

A poor body condition score at calving adversely affects fertility, characterized by prolonged post-partum intervals, reduced conception rates, and more services per conception. A very low protein diet can cause cessation of estrus. (Agrawal *et al.*, 2003).

Approaches for improving reproductive efficiency:

During the last two decades, farm animal reproduction has entered into the area of a new biotechnology revolution which includes artificial insemination; induction and synchronization of estrus (manipulation of breeding cycle); super ovulation embryo transfer; cryopreservation; embryo resource development; sexing; transgenesis; cloning chimera production; and early pregnancy diagnosis. Some of these technologies have immense potential to revolutionize world animal agriculture in the twenty-first century.

Artificial insemination: One effective ways of rapidly accelerating significant genetic improvement is A.I., but A.I. is not successfully practiced in buffalo because

1. Difficulties in estrous detection.
2. Difficulties in semen preservation.
3. Lack of certified semen programmes.

Reproductive performance of buffaloes:

Reproductive efficiency is determined by many different processes, which result from interaction among genetic and environmental factors. The processes involved singly or in concert, include age of puberty/maturity, pattern of estrous cycle and estrous behaviour, length of breeding, ovulation rate/litter size, lactational anoestrous period, post-partum anoestrus intercalving period, and reproductive life span. These traits combined measure breeding efficiency/performance (Agrawal, 2003).

The reproductive effciency in buffalo is so alarmingly low that it poses a very serious threat of economic loss to Indian and animal husbandry professionals (Ramesh *et al.*, 2002). In such a scenario, an ample scope exists for increasing the reproductive efficiency by modification in the traditional methods of breeding, feeding, management and disease control.

Reasons for poor reproductive performance:

1. Effect of climate: A determining factor in production and reproduction of farm animals all over the world is environment. Season affects the breeding efficiency in the case of buffaloes. There is a tendency to have better performance during the cool months; 70-80% of conceptions in buffaloes occursbetween July and February. Buffaloes are sexually activated by decreased day length and temperature. A lower number of services per conception are needed during the July-February breeding season than the March-June breeding season (Agrawal *et al.*, 2003).
2. Poor thermal tolerance: Buffaloes have poorly developed thermal regulation system. During the summer they have to be protected from the extreme heat by allowing them wallowing; also in winter, they have to be protected from extreme cold, which may predispose them to many diseases. (Ramesh *et al.*, 2002).

One of greatest source of failure in A. I. is the inability to recognize estrous display. With frozen semen, the need to inseminate close to ovulation is even more imperative because of the limited survival of spermatozoa. It is possible to extend the life of spermatozoa and package them in such a way as to retain them in the reproductive tract in a viable state for several days. Efforts are being made for micro-incapsulation of spermatozoa, which will help in survival of spermatozoa.

There is an urgent need to start registration of all A.I. bulls by a national society or an all India animal breeder association and to initiate a certified semen programme. Strengthening practical training in animal reproduction for veterinary students and better training of inseminators is also essential. (Agrawal*et al.*, 2003).

1. Insemination should be done during the latter half of the period of heat.
2. In heifers, the first heat should be avoid to make the animal more receptive.
3. Buffalo should be calm and quiet at time of insemination
4. The Best site for insemination is the mid cervix and the body of the uterus (Ramesh *et al.*, 2002).

2. Induction and synchronization of estrous:

Controlling the estrous cycle and regularizing it. A. I. would provide a means for circumventing the problem of estrous detection. For a satisfactory pregnancy rate in an embryo transfer programme, the embryo must be placed in an environment that simulates the one from which it was removed. So synchronization of estrous in the donor and recipient in each experiment is considered essential to obtain the best result.

Induction and synchronization of estrous can be achieved by two alternate approaches. The first is by artificially extending the luteal phase (by using progestatimal compound) and second by inducing device of the corpus luteum (by using prostaglandins and analogues.).

Progestogens with some possible adjustment in PMSG and or FSH are used for anticipated lower response. Presently PGF2 a is widely used to manipulate the early breeding of post-partum buffaloes and pubertal heifers. It is also used in a double dose schedule (10-11 days interval) for those animals, which do not have functional CL at the first injection. (Agrawal *et al.,* 2003).

Superovulation (multiple ovulation): Superovulation is one of the major reproductive technologies for rapid genetic improvement of livestock. (Agrawal *et al.,* 2003).

Reduction of unproductive period: Method of reducing the interval from calving to conception and decreasing no. of services per conception. (Ramesh *et al.,* 2002).

Oral Prostagen (Chlormadinone acetate, melangestrol acetate) for a 14-day period, starting on the 21st day after calving was found to be effective in reducing interval between parturition and conception.

Short-term progesterone (PRID/ CIDR) and progestagen (ear implant) for 7-10 days also stimulates breeding activity regardless of whether the animal is cyclic or anoestrous. FSH in conjugation with progesterone enhanced the follicular development in the post-partum period. Progesterone plus estradioltreatment resulted in synchronized wave emergence with normal luteal activity in post-partum anoestrous animals. (Ramesh *et al.* 2002).

Embryo transfer technique: The embryo transfer technique permits exploitation of superior female genotype, giving more off-spring from the same genetic donor than would arise under normal conditions of breeding. Embryo technologies are also used to resolve several reproductive enigmas, viz. uterine sufficiency; maternal reorganization of pregnancy; embryo-utero-relationship. (Agrawal *et al.,* 2003).

6

Assisted Reproductive Technologies

Artificial insemination (AI) has been used to obtain offspring from genetically superior males for more than 200 years. Improvements in methods to cryopreserve (freeze) and store semen have made AI accessible to more livestock producers. In the same manner as cryopreservation of semen, embryo freezing allowed for the global commercialization of animals with high genetic qualities. Semen from bulls is especially amenable to freezing and long-term storage. In the dairy industry, where large numbers of dairy cows are managed intensely, AI is simple, economical, and successful. More than 60 percent of dairy cows in the United States are bred by AI. However, the situation is different for beef cattle, where breeding populations are usually maintained on range or pasture conditions. In the United States beef industry, AI accounts for less than 5 percent of inseminations. For reasons that are not yet well understood, it is more difficult to freeze and store semen from other livestock species, including horses, pigs, and poultry, than it is to freeze cattle semen. NIFA has supported research projects to understand the physiological processes associated with cryopreservation of semen or embryos and to develop improved methods of cryopreservation for gametes (eggs and sperm) and embryos from several livestock species.

MULTIPLE OVULATION AND EMBRYO TRANSFER

Development of embryo transfer technology allows producers to obtain multiple progeny from genetically superior females. Depending on the species, fertilized embryos can be recovered from females (also called embryo donors) of superior genetic merit by surgical or nonsurgical techniques. The genetically superior embryos are then transferred to females (also called embryo recipients) of lesser genetic merit. In cattle and horses, efficient techniques recover fertilized embryos without surgery, but only one or sometimes two embryos are produced during each normal reproductive cycle. In swine and sheep, embryos must be recovered by surgical techniques. To increase the number of embryos that can be recovered from genetically superior females, the embryo donor is treated with a hormone regimen to induce multiple ovulations, or superovulation. NIFA has supported basic and applied research to enhance the efficiency of superovulation and embryo transfer in livestock species.

IN VITRO FERTILIZATION

As an alternative to collecting embryos from donor animals, methods have been developed recently to produce embryos in vitro (in the laboratory). The methods are also called in vitro embryo production. Immature oocytes (female eggs) can be obtained from ovaries of infertile or aged females, or from regular embryo donors (described above). Ovum (egg) pick up is a nonsurgical technique that uses ultrasound and a guided needle to aspirate immature oocytes from the ovaries. Once the immature oocytes have been removed from the ovary, they are matured, fertilized, and cultured in vitro for up to seven days until they develop to a stage that is suitable for transfer or freezing. NIFA has supported a number of basic research projects to understand the physiological mechanisms associated with production of embryos in vitro.

SEX DETERMINATION OF SPERM OR EMBRYOS

The beef industry in the United States prefers male calves, which tend to have higher body weights and higher feed efficiency (compared to female or heifer calves) when placed in feedlots for the growing and finishing stages of meat production. In contrast, the dairy industry prefers heifer calves, which will ultimately produce offspring and milk for human consumption. Thus, methods are needed to determine the sex of sperm or embryos so producers can control the sex of the offspring of their livestock. Using a specific dye that binds to DNA (the Hoechst 33342 stain) and a flow cytometer/cell sorter, the DNA content of individual sperm is measured. In cattle, the X-bearing sperm contain 3.8 percent more DNA than the Y-bearing sperm. In mammals, the presence of a Y chromosome (and one X chromosome) determines that the individual will be a male. Female mammals contain 2 X chromosomes. Although the process to sort the X and Y bearing sperm is slow (approximately 10 million live sperm of each sex can be obtained per hour—this is about the number of live sperm required for one conventional dose of frozen semen for artificial insemination), this procedure determines the sex with higher than 95 percent accuracy.

NUCLEAR TRANSFER OR CLONING

Since the mid 1980s, technology has been developed to transfer the nucleus from either a blastomere (cells from early, and presumably undifferentiated cleavage stage embryos) or a somatic cell (fibroblast, skin, heart, nerve, or other body cell) to an enucleated oocyte (unfertilized female egg cell with the nucleus removed). This "nuclear transfer" produces multiple copies of animals that are themselves nearly identical copies of other animals (transgenic animals, genetically superior animals, or animals that produce high quantities of milk or have some other desirable trait, etc.). This process is also referred to as cloning. To date, somatic cell nuclear transfer has been used to clone cattle, sheep, pigs, goats, horses, mules, cats, rabbits, rats, and mice.

The technique involves culturing somatic cells from an appropriate tissue (fibroblasts) from the animal to be cloned. Nuclei from the cultured somatic cells are then microinjected into an enucleated oocyte obtained from another individual of the same or a closely related species. Through a process that is not yet understood, the nucleus from the somatic cell is reprogrammed to a pattern of gene expression suitable for directing normal development of the embryo. After further culture and development in vitro, the embryos are transferred to a recipient female and ultimately result in the birth of live offspring. The success rate for propagating animals by nuclear transfer is often less than 10 percent and depends on many factors, including the species, source of the recipient ova, cell type of the donor nuclei, treatment of donor cells prior to nuclear transfer, the techniques used for nuclear transfer, etc. NIFA has supported research projects to obtain a better understanding of the basic cellular mechanisms associated with nuclear reprogramming.

CURRENT AND FUTURE ASSISTED REPRODUCTIVE TECHNOLOGIES FOR MAMMALIAN FARM ANIMALS

Reproduction in domestic animals is under control by man and the technologies developed to facilitate that control have a major impact on the efficiency of food production. Reproduction is an energy-intensive process. In beef cattle, for example, over 50 % of the total feed consumption required to produce a unit of meat protein is consumed by the dam of the meat animal (Anim Prod 27:367-379, 1978). Sows are responsible for about 20 % of the total feed needed to produce animals for slaughter (Adv Pork Prod 19:223-237, 2008). Accordingly, energy input to produce food from animal sources is reduced by increasing number of offspring per unit time a breeding female is in the herd. Using beef cattle as an example again, life-cycle efficiency for production of weaned calves is positively related to early age at puberty and short calving intervals (J Anim Sci 57:852-866, 1983). Reproductive technologies also dictate the strategies that can be used to select animals genetically for traits that improve production. Of critical importance has been artificial insemination (AI) (Anim Reprod Sci 62:143-172, 2000; Stud Hist Philos Biol Biomed Sci 38:411-441, 2007; Reprod Domest Anim 43:379-385, 2008; J Dairy Sci 92:5814-5833, 2009) and, as will be outlined in this chapter, emerging technologies offer additional opportunities for improvements in genetic selection. Given the central role of reproduction as a determinant of production efficiency and in genetic selection, improvements in reproductive technologies will be crucial to meeting the challenges created by the anticipated increases in world population (from seven billion people in 2011 to an anticipated nine billion by 2050; World population prospects: the 2010 revision, highlights and advance tables. Working Paper No. ESA/P/WP.220, New York) and by difficulties in livestock production wrought by climate change (SAT eJournal 4:1-23, 2007).The purpose of this chapter will be to highlight current and emerging reproductive technologies that have the potential to improve efficiency of livestock production. The focus will be on technologies that manipulate male and female gametes as well as the stem cells from which they are derived and the preimplantation embryo. While technology is crucial to other interventions in the reproductive process like control of seasonal breeding, hormonal regulation of ovulation, estrous cyclicity and pregnancy establishment, feeding to optimize reproduction, minimizing environmental stress, and selection of genes controlling reproduction, these will not be considered here. Rather the reader is directed to other chapters in this volume as well as some reviews on other aspects of artificial manipulation of reproduction (Reprod Fertil Dev 24:258-266, 2011; Reprod Domest Anim 43:40-47, 2008; Reprod Domest Anim 43:122-128, 2008; Soc Reprod Fertil Suppl 66:87-102, 2009; Comprehensive biotechnology, Amsterdam, pp 477-485; Dairy production medicine, Chichester, pp 153-163; Theriogenology 76:1619-1631, 2011; Theriogenology 76:1568-1582, 2011; Theriogenology 77:1-11, 2012). Given the large number of mammalian species used for production of products useful for man and the diversity in their biology and management, the review will not be comprehensive but instead will use results from species that are most illustrative of the opportunities generated by assisted reproductive technologies.

Assisted reproductive technologies used in domestic species

Assisted reproductive technology (ART) is the term used to describe treatments and procedures in the field of reproductive biology used to establish pregnancy by artificial or partially artificial means. This usually involves the hormonal manipulation of reproductive cycles as well as the utilisation of gametes or embryos, including artificial insemination, induction of ovulation, in vitro fertilisation, gamete and embryo cryopreservation and similar procedures designed to enhance pregnancy in treated animals.

The goals of ARTs vary depending on the species or groups of animals to which they are applied. In humans, ARTs are used almost exclusively to overcome issues of either male or female infertility or, in some cases, for pre implantation diagnosis of genetic disorders. In rodents such as mice, they are used as a valuable research model to facilitate better understanding of the underlying regulation of normal and abnormal developmental processes; for example, the production of 'knockout' mice has provided an extremely informative research tool in the study of gene function. Initially, the focus of ARTs in domestic species was concerned with genetic improvement or production aimed at allowing the livestock industry to respond to continually increasing demands for improved productivity and quality (for example, producing more offspring from a valuable animal than would be possible through normal mating), as well as the conservation of rare or endangered breeds or species. In recent years, however, the use of certain techniques, such as somatic cell nuclear transfer (SCNT) and genome editing has opened up applications in the biomedical model area.

Several generations of ARTs have been developed for domestic animals. These include artificial insemination (AI), multiple ovulation embryo transfer (MOET), in vitro embryo production (IVF) and related techniques including transvaginal oocyte recovery (often referred to as 'ovum pick up' or OPU), cloning and transgenesis. The main impetus for the development of these technologies arose from their potential application in the refinement of breeding strategies for improved production and health in animal husbandry. More recently, biomedical applications of these technologies, in particular, somatic cell nuclear transfer and stem cell culture, have been pursued in domestic mammals as possible models applicable to human clinical conditions or with potential application in therapeutic drug development.

As there are major differences in reproductive physiology between mammalian species, it is not surprising that reproductive technologies can be highly efficient in some animal groups, including most farm animals and laboratory rodents, but are very inefficient when used with carnivores, which often have species specific reproductive features. Among domestic species, the application of ARTs has been extensively employed in cattle, which reflects their economic importance in food production. Accordingly, in this chapter, emphasis will be placed on the use of ARTs in cattle followed by a discussion of their application in other species.

The impact of reproductive technologies on animal breeding

The rate of genetic improvement in most breeding programmes is controlled by four main factors: (1) the selection intensity, a measure of how selective breeders are in the choice of animal; the fewer animals selected based on their superior performance for a particular trait, the faster the rate of genetic progress; (2) the accuracy with which the genetic merit of an individual animal can be predicted; the greater the accuracy, the greater the potential improvement; (3) the genetic variation in the particular trait in question; the greater the variation for a given trait, the greater the scope breeders have to select animals which are far from the mean level of performance for the trait; and (4) the generation interval, a measure of how long it takes the selected animal(s) to contribute their superior genes to the next generation. In domestic species, this tends to be quite a long period, up to three years. Coupled with these four parameters are the selection differential (the difference in performance in a particular trait between the selected animals and the overall group from which they are selected) and the heritability of the trait (the proportion of the selection differential which, on average, is passed on to the offspring).

Several reproductive biotechnologies can influence one or more of these parameters and, thereby, enhance the rate of genetic improvement (Fig 27.1). Advantages of MOET, for example, include higher female selection intensity and increased selection accuracy. However, variability in an individual animal's response to superovulation and the low average number of transferable embryos recovered are still limiting factors. In vitro embryo production can potentially affect generation interval through its use on prepubertal donors. Indeed, with the recent advent of genomic selection, it is now possible to select embryos based on a biopsy taken a few days after fertilisation.

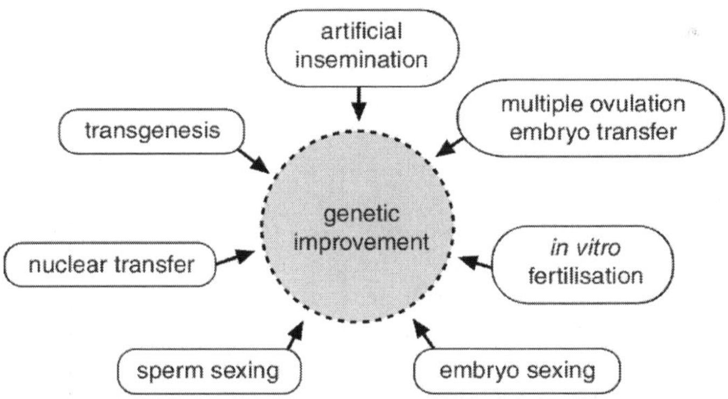

Reproductive technologies used to enhance genetic improvement.

Assisted reproductive technologies in cattle

Artificial insemination

Among the available reproductive technologies (Fig 27.2), artificial insemination (AI) has made the single biggest impact on animal production and genetic improvement. Artificial insemination allows high selection intensity on the male side, permitting superior males to sire in excess of 100,000 offspring in their lifetime. At each ejaculation, a bull deposits up to 10 billion spermatozoa in the cranial vagina of the cow. All of these spermatozoa can potentially compete to fertilise the single oocyte that is typically ovulated and only one spermatozoon is required for fertilisation. Indeed, penetration by more than one spermatozoon results in polyspermy, leading to an abnormal chromosome number and embryo mortality. In reality, however, the vast majority of spermatozoa never reach the site of fertilisation in the uterine tube. Thus, there is substantial wastage associated with natural service. In contrast, typical semen doses used in conventional bovine AI, where semen is placed directly into the uterus, contain between 10 and 20 million spermatozoa when frozen thawed semen is used. Furthermore, when fresh or sex sorted spermatozoa are employed, concentrations as low as one to three million spermatozoa are used. Depending on volume and spermatozoa concentration, a single ejaculate can potentially be used to inseminate more than 1000 cows.

MOET: multiple ovulation embryo transfer; AI: artificial insemination; IVF: in vitro fertilisation; NT: nuclear transfer.

Advantages of AI include:

- allowing the widespread use of the semen of outstanding sires;
- storage of semen for future use, even after the death of the sire;
- accelerating the introduction of improved genetic characteristics;
- flexibility in the choice of sire compared to natural service (for example, on the basis of breed, calving ease, production traits and other desirable characteristics);
- optimising the cost effectiveness – bulls are expensive to buy and rear, without the assurance of high fertility;
- minimising the risk of disease transmission compared to natural mating;
- reducing the risk of injury to humans and animals;
- facilitating the use of oestrous synchronisation and sex sorted semen.

Multiple ovulation embryo transfer (MOET)

High genetic merit sires produce billions of fertile gametes at each ejaculation and can sire thousands of offspring during, or even after, their lifetime. However, the contribution of a genetically superior cow to a breeding programme is limited by the fact that cows are monovulatory (that is, only one oocyte is ovulated during each oestrous cycle) and gestation lasts nine months following which a period of uterine involution is required before another pregnancy can be initiated. In addition, these animals have a relatively short productive lifespan of approximately five years.

Induction of multiple ovulations (MO) in a donor, often referred to as 'superovulation', coupled with AI, embryo recovery and embryo transfer (ET) to surrogate recipients, provides an opportunity to increase the impact of superior females on a breeding programme, comparable to the impact of AI on the role of the male, but on a much smaller scale.

As the name suggests, MOET involves several steps including (1) synchronisation of the oestrous cycles of the donor and recipient females, (2) induction of superovulation in the donor, (3) AI of the donor, (4) recovery of embryos from the donor, and (5) transfer of embryos to recipients or cryopreservation of the embryos for future transfer.

IN VITRO FERTILISATION (IVF)

The ability to produce embryos in vitro (in vitro embryo production, IVP) has been possible for more than three decades in domestic mammals, particularly cattle.

The first human conceived following IVF, Louise Brown, was born in 1978. The first calf produced by IVF was born in 1981. In both instances, the procedure involved the removal of a matured ovulated oocyte from the uterine tube, incubation with spermatozoa in vitro and subsequent replacement in the female reproductive tract.

The in vitro production of bovine embryos is a three step process involving oocyte maturation (in vitro maturation, IVM), oocyte fertilisation (IVF) and embryo culture (IVC). Despite these three distinct steps, the term 'IVF' is often used generically to refer to the entire process. Immature oocytes are typically collected either from the ovaries of slaughtered heifers and cows at an abattoir or from the ovaries of live animals using transvaginal oocyte recovery procedures. Good quality cumulus oocyte complexes are selected, based on their morphological appearance, and are matured in an incubator for 24 hours, typically in an appropriate culture medium supplemented with gonadotrophins and/or growth factors. The first morphological sign that maturation has occurred is the expansion of cumulus cells surrounding the oocyte. Removal of the cumulus cells and more detailed examination of the oocyte itself reveals the first polar body extruded into the perivitelline space, indicating attainment of metaphase II, the stage at which the oocyte is normally ovulated in vivo and at which fertilisation occurs. For IVF, oocytes are incubated in vitro with fresh or, more typically, frozen thawed spermatozoa. Prior to addition, motile spermatozoa are isolated, typically using density gradient centrifugation. Gametes are co incubated for up to 24 hours after which the presumptive zygotes are washed and cultured for seven days in an appropriate medium to the blastocyst stage, when they are transferred to a recipient or cryopreserved for subsequent transfer.

Overview of multiple ovulation and embryo transfer in cattle. Following superovulation, the high genetic merit donor is inseminated with semen from a high genetic merit sire. Approximately seven days later, embryos are recovered by non surgical uterine flushing. Good quality embryos are transferred to synchronised recipients. Meanwhile, the valuable donor can be resynchronised and superovulated to produce more embryos, or she can be returned to natural breeding.

Sir Walter Heape is credited with carrying out the first embryo transfer in 1890 using rabbits. The first successful embryo transfer in cattle was recorded in 1951. Since that time, this procedure has played an important role in cattle breeding worldwide. Historically, embryo recovery and transfer involved the use of surgical procedures. However, the development of simple non surgical recovery and transfer procedures in the 1970s ensured more widespread accessibility of the technology. Data collated annually by the International Embryo Technology Society indicate that approximately one million cattle embryos are transferred worldwide each year.

Superovulation protocols have become more refined over the past 50 years. The use of commercial pituitary extracts and prostaglandins in the 1970s, and partially purified pituitary extracts and progesterone releasing devices in the 1980s and 1990s, have facilitated the development of many of the protocols used today. Furthermore, knowledge of follicular wave dynamics through the use of real time ultrasonography and the development of strategies for control of follicular waves have provided new practical approaches to superovulation. Despite the fact that much research has focused on methods to maximise the number of ovulations in the donor female, the total yield of transferable embryos has not changed markedly over the last 40 years (typically about five to seven transferable embryos are recovered per superovulated donor). Variability in the donor's response to superovulation continues to be one of the main limitations of embryo transfer in cattle.

Many of the benefits or potential applications of AI which also apply to ET include:

- multiplication of genetically superior females;
- long distance/international transport of embryos without the necessity of acclimatisation and health issues associated with live animal transportation;
- long term storage of embryos;
- increased twinning rates in cattle, either by the transfer of multiple embryos or the transfer of a single embryo to the uterine horn contralateral to the corpus luteum of a previously inseminated recipient;
- providing twins for research, thereby reducing the numbers of animals required;
- overcoming the negative effects of heat stress, particularly in lactating dairy cows, where the quality of the oocyte can be compromised;
- facilitating the control of the sex of a calf (by sexing the embryo prior to transfer or by predetermining the sex of the embryo using sex sorted spermatozoa);
- conservation of endangered species;
- maximising the number of calves per semen dose when used with superovulated donors;
- minimising the risk of transmission of infectious agents compared with natural service.

Technology Overview: Somatic Cell Nuclear Transfer and Other Assisted Reproductive Technologies

Since the beginnings of livestock agriculture, selection criteria have been applied to foster the propagation of animals with traits more desirable to humans. The expansion of herds with desirable traits has been limited, however, by the reproductive capacity of the species or breed and the prevalence of particular versions of genes (or sets of genes) responsible for those traits in the available gene pool. (The gene pool can be considered all of the animals available for breeding.) The female contribution to reproductive success, for example, is limited by species-specific characteristics such as average litter size, frequency of estrus, and gestation length. In natural breeding, male contributions are restricted by the degree of proximity to fertile females and the ability to inseminate females with a sufficient number of normal sperm. Finally, individuals of both sexes are limited to the extent that they may carry the desired versions of genes or combination of genes.

To help overcome some of these complications, various forms of assisted reproductive technologies (ARTs) have been adopted in animal agriculture for over a century, and at least one (artificial insemination) has been used for several hundred years. These technologies form a continuum that ranges from the fairly minimal assistance provided to animals engaged in natural service through those containing components of significant in vitro manipulation such as in vitro fertilization and embryo splitting, to the more recent development of somatic cell nuclear transfer (SCNT), or what is colloquially referred to as "cloning"[8]

(Faber et al. 2004; Sakai 2005; Oback and Wells 2007b). Beginning with the development and application of modern artificial insemination (AI) methodologies in the first half of the 20th century, ARTs have aided in the genetic improvement of domestic species, including selection of phenotypes such as behavioral and production traits in domesticated animals (Youngquist 1997, Faber and Ferre 2004). By accelerating the rate at which selective breeding goals can be met, improved genotypes have expanded rapidly into national herds in the United States and other countries (Faber et al. 2004; Wells 2005). In turn, this has resulted in lower costs for livestock producers and retail consumers, while simultaneously maintaining or improving the quality and consistency of foods of animal origin.

Reproductive technology advances have also proven to be powerful tools in curbing the spread of vertically transmitted diseases (i.e., those that are passed from the dam to her offspring during the period immediately before and after birth, either across the placenta or in the dam's milk) (Youngquist 1997). For example, embryo transfer (ET) (see subsequent discussion for a description of this ART) has been used to prevent vertical transmission of Neospora caninum in cattle (Baillargeon et al. 2001; Landmann et al. 2002), scrapie in sheep (Wang et al. 2001), Bovine Virus Diarrhea (BVD) in cattle (Smith and Grimmer 2000), and Brucella abortus in an American bison (Robison et al. 1998). Embryo transfer is commonly used in laboratory animal research to re-derive valuable strains of gnotobiotic (i.e., animals in which all of the bacterial species are known) or specific pathogen-free research animals when colonies become infected with undesirable disease agents that cannot be controlled through more conservative means.

The following chapter begins with a brief overview of what cloning is, followed by an overview of the continuum of other ARTs commonly in use in current US agricultural practice, placing nuclear transfer technology into context of these breeding practices. Appendix B provides additional details on overall reproductive efficiency observed in current agricultural practice in the US, and Appendix C provides a comprehensive summary of the outcomes observed in ARTs, with particular emphasis on those technologies that contain a significant in vitro culturing component. Although all of these technologies are currently in practice, all are continually undergoing development and refinements with the goal of improving efficiencies. A reasonable expectation then, is that success rates (defined as the rate of production of healthy animals) will improve as expertise increases.

A. What is Cloning?

Cloning, or somatic cell nuclear transfer, is a process by which animals are reproduced asexually (embryo splitting and blastomere nuclear transfer are other ways of reproducing animals asexually and are discussed later in this chapter). In cloning, a differentiated somatic cell (a non-germ line cell from an existing animal) is introduced to an oöcyte (a cell that is the immediate precursor of a mature egg) that has had its nucleus (and thus its genome) removed, and then, following some manipulations, is induced to start replicating. If all goes well, the dividing cell is implanted into a female animal (dam), continues to develop normally, and is delivered just as any newborn.

Since the first report of a clone produced by SCNT (Wilmut et al. 1997), several other species have been cloned (Table II-1), although in some cases (e.g., companion animals) only a limited number of animals have been generated. The reasons for this are multi-fold, but are largely driven by the relative difficulty in producing clones, and the various drivers, economic and technical, that affect the expansion of the technology. For example, the use of clones in expanding elite breeding stock in domestic livestock is perceived to have benefit for breeders and consumers. This risk assessment does not attempt to address those issues, however, and instead concentrates on those domestic livestock clones commonly consumed as food (e.g., cattle, swine, sheep, and goats).

B. Continuum of Reproductive Technologies

Natural Service

Although many people who are not involved in intensive animal agriculture assume that most breeding occurs "naturally"[9] (e.g., a male animal mates with receptive female), in fact, human intervention is the industry standard for many livestock operations (Youngquist 1997). In the US dairy industry, for example, most reproduction involves some technological component, and swine producers rarely use natural mating for their production of offspring. Conversely, in the beef industry most reproduction occurs by natural service, and most of the world's sheep and goat production occurs under free range conditions and depends on natural mating.

Humans have assisted animals in natural mating by monitoring the reproductive status of females, introducing receptive females to the same location (e.g., field, corral, or pen) as the male, and allowing nature to take its course. When this process does not result in sufficient offspring of the desired phenotype, or is otherwise compromised, assisted reproductive technologies can be called into play.

ARTIFICIAL INSEMINATION AND SYNCHRONIZED ESTRUS

The first ART developed was artificial insemination (AI), which in its simplest form involves the collection of semen from males and its subsequent human-assisted introduction into a physiologically receptive female. It is an important technique for the genetic improvement of animals, as a few select males can produce sufficient sperm to inseminate thousands of females per year, while natural service would provide for the insemination of only a fraction of those animals.

Reports of AI in horses as part of breeding programs have been traced to the Arabian Peninsula in the 14th century (Bearden and Fuquay 2000). AI of a beagle dog was first described by Spallanzani in 1780 (Hafez and Hafez 2000). In 1899, the Russian Czar Nicholas II commissioned I.I. Ivanov to develop an AI program for horses, and by 1933 Ivanov had developed methods for collecting semen and inseminating horses, cows, sheep, and pigs (Foote 2001). In 1931, 19,800 cows were bred by AI in Russia. By 1936, Denmark had established an AI cooperative association, and by 1939, the use of AI had spread to the United States. In 1970, it was estimated that 7,344,420 dairy cows were bred using AI (Webb 2003).

Although there are several methods for collecting semen, most involve training males to ejaculate into an artificial vagina. Semen is then diluted to maximize the number of services that one male can provide. A normal ejaculate from a dairy bull usually contains between 5 and 10 billion sperm; good conception rates generally require about 12-20 million sperm to be introduced. The diluting solution contains factors that help to stabilize and preserve the sperm, as well as antibiotics to inhibit bacterial growth and reduce the danger of spreading any potential disease or contamination. Most collected semen is stored in glass ampoules or plastic straws, and is generally stored either in dry ice and alcohol (-100oF) or liquid nitrogen (-320oF). To date, there appears to be no limit on the amount of time that bovine sperm can remain frozen and regain viability upon appropriate thawing. Since 1997, use of AI in swine breeding has increased dramatically. A survey of swine producers conducted by the National Pork Board in 2003 indicated that even among small producers (1,000 to 3,000 swine marketed annually) as many as 60 percent of litters were sired by AI in 2003, while for large producers (> 50,000 swine marketed annually) 98 to 100 percent of litters were sired by AI.[10]

Rams (male sheep) and bucks (male goats) can also be donors for artificial insemination.

In the US, AI of the female is usually performed either by trained technicians employed by breeding companies or large farms or by the producers themselves. The most common technique employed today for dairy cows involves the use of sterile, disposable catheters that are inserted vaginally and extended through the cervix into the body of the uterus of the recipient cow (whose estrous cycle has been documented). Thawed semen is warmed to the appropriate temperature, and sperm are deposited in the uterine/cervical regions.

The primary advantages of AI to farmers include the ability to use semen from bulls anywhere in the world rather than those that are more geographically proximate, and thus to have desirable genetics available for propagation. It also allows the farmer to use multiple sires in a herd without the attendant costs of maintaining animals that are often difficult to handle and in multiple breeding pastures. AI tends to be less expensive than natural service (a straw of semen generally costs less than transporting a female to the sire and the stud fee) and avoids the potential physical risks to either sire or dam as part of the mating process. The disadvantages of AI include the need to train personnel engaged in the breeding operations on how to detect estrus in females (see subsequent discussion of estrous synchronization), and training or retaining individuals to perform the insemination. Further, care needs to be taken not to rely excessively on a few apparently superior sires so as not to reduce the genetic diversity of the resulting herds.

Sperm collection and AI were further improved by the advent of sperm sexing, or selection of sperm carrying an X (female) or Y (male) chromosome.[11] Development of an effective and simple method for producing animals of the desired sex is economically desirable for livestock producers; sperm sexing is currently being used when available and economically feasible (Foote 2001; Faber and Ferre 2004). For example, in the dairy industry, females are desired because males do not produce milk; and excess males often become veal. In the beef industry, however, males are desired because they grow faster. Females can be the desired sex in the swine industry where leaner animals generally receive higher prices; young female pigs (gilts) tend to be leaner than castrated male pigs (barrows) when they arrive at market.

One method that shows the most promise for predetermining the sex of offspring is sexing semen using flow cytometry. This technique is based on the observation that in livestock species, sperm with X chromosomes have about 3 percent more DNA than those with Y chromosomes. Collected semen is diluted, and single sperm are passed through a laser beam that allows for the determination of the amount of DNA in each individual sperm. Based on their relative DNA content, sperm are sorted into "heavier" (female producing) and "lighter" (male producing) fractions. Another method sexes early embryos by removing one or two of the cells from the early embryo, arresting the further growth of the embryo by freezing, and identifying genes found only on the Y chromosome using the polymerase chain reaction (PCR) in the selected cells (Youngquist 1997). Semen sexing is more rapid, less invasive, and more economical, while embryo sexing is impractical at this time, as it is invasive, time intensive, and quite expensive. Further, the potential to damage the embryo by piercing the protective layer around it (zona pellucida), removing cells, and freezing the remaining cells in the embryo is quite high.

Estrous synchronization, or the timed induction of estrus (commonly referred to as heat), further permits the efficient use of artificial insemination (Hafez and Hafez 2000). Estrous synchronization was first practiced in the US in the 1960s and allows for the insemination of large groups of animals in a specified window of time when the female is capable of conceiving a pregnancy. The alternative to estrous synchronization is the time- and labor-intensive method of observing females' behavior to detect estrus, and then arranging insemination at the appropriate time on an animal-by-animal basis. Estrous synchronization is achieved by treating females with hormones (Hafez and Hafez 2000). These synchronization regimens mimic the natural hormonal patterns leading to estrus and ovulation, and allow the producer to control the timing of these patterns. Several FDA-approved hormones are available for estrous synchronization, such as progesterone and prostaglandin F2a.[12] These hormones are commonly used for conventional breeding in cattle throughout the U.S. According to a recent survey of large U.S. commercial dairies, 87 percent of herds use hormonal estrous synchronization methods (Caraviello et al. 2006). The advantage of linking AI to estrous synchronization lays in the ability of contained agricultural practices to operate on a more predictable schedule. For example, cattle breeders can avoid the reduced conception rates that occur during summer's heat by breeding animals during the cooler spring season. Predictability can benefit farmers by allowing them to allocate resources (e.g., farm labor, veterinary visits) more efficiently, thus lowering production costs.

Embryo Transfer

It is impossible for a fertile female mammal to bear all of her potential offspring. Litter size, gestation time, and post-partum decreases in fertility all limit the potential number of progeny that she can produce. When the female animal reaches the end of her reproductive period, any remaining unfertilized eggs represent potential offspring that have been lost. One solution to this dilemma is to transfer embryos of genetically superior female animals to multiple surrogate dams. This technique, called "embryo transfer," is particularly useful in species in which a low number of progeny are produced per gestation.[13]

In concept, then, embryo transfer (ET) is analogous to AI in that the total yield of offspring from a genetically superior, in this case, female animal can be increased (Youngquist 1997).

In 1890, rabbit embryos were first transferred from a donor female to surrogate rabbits. The experiment demonstrated that the surrogate's genetics would not influence the transferred embryo's genetics or development. In 1951, a successful live bovine ET was accomplished, but non-surgical methods of embryo collection did not succeed until the late 1960s (Hafez and Hafez 2000).

Currently, it is possible to flush large numbers of viable embryos from a superovulated cow with minimal stress to the animal (Hafez and Hafez 2000). Superovulation of the donor animal is generally accomplished by injecting the animal with follicle stimulating hormone or other exogenous gonadotropins before she enters estrus. The hormones induce production of a large quantity of ovarian follicles containing mature, preovulatory oöcytes. Insemination is performed at appropriate times relative to ovulation depending on the species and breed. Using standard synchronization methods, the estrous cycle in recipient surrogate mothers is synchronized in parallel with the cycle of the donor to be ready to accept embryos for implantation and gestation. When embryos are about a week old, they may be flushed out of the donor dam's uterus nonsurgically[14] by guiding a catheter through the cervix and into the uterus. Saline solution is injected into the uterus via the catheter and then flushed, along with the embryos, back out of catheter into a collection vessel. Embryos are isolated from the flushing solution, and examined microscopically to determine whether they are of sufficiently quality to implant. If they meet the criteria for further use, embryos can be transferred immediately to a waiting synchronized recipient animal using the same non-surgical technique that was used to collect the embryos. Alternatively, embryos can be frozen for later use or split into halves (see embryo splitting discussion below). Fresh or thawed embryos are inserted into surrogate mothers, where they attach to the lining of the uterus. Following transfer of embryos, no additional hormones or special treatments are required and the pregnancy is allowed to progress normally.

This technique, referred to as MOET (multiple ovulation and embryo transfer), is often used in relatively intensive cattle breeding programs, but is less developed in other livestock species. Similar to fertilizing many females with sperm from one superior male, MOET provides the breeder the ability to expand genetic traits exhibited in superior females. Further, the ability to freeze embryos allows for the preservation of "genetic stock" to be used at a later time. Its prevalence in livestock breeding, however, is much lower than AI, as it is considerably more expensive (Wilmut et al. 2002).

The International Embryo Transfer Society (IETS), a professional society whose membership includes breeders and researchers, estimates that a total of approximately 612,178 in vivo derived bovine embryos were transferred worldwide in 2005 (Thibier 2006). This is an increase of approximately 62,000 embryos from the previous year. Most of those transfers occurred in North America (45.2 percent), with the rest taking place in South America (20.5 percent), Asia (18.9 percent), and Europe (~14 percent).

In vitro Fertilization

The first in vitro fertilized (IVF) offspring was a rabbit born in 1959 (Chang 1959a). Since that time, IVF offspring have been born to mice, rats, hamsters, cats, guinea pigs, squirrels, pigs, cows, monkeys, and humans (Bearden and Fuquay 2000). IVF allows for the production of offspring from animals where other ART methods fail due to difficulties with either the female (blocked oviducts, non-responsive ovaries) or male (marginal semen quality and/or quantity), or where disease is present. In cattle, it is also used for the production of embryos from sexed semen because of the low sperm counts resulting from current sexing protocols, and for the further extension of the semen of superior sires due to the relatively low level of sperm required for in vitro fertilization. (IVF procedures are also used to assist human couples with limited fertility.)

The overall technique for IVF is similar among species, and involves significant manipulations in vitro, or outside the body of animals. In livestock species, oöcytes are collected from the ovaries of either living or deceased animals whose genetic potential is desirable (Goodhand et al. 1999). Ovaries can be obtained by transvaginal aspiration from live animals, or from a deceased animal at time of slaughter. Slaughterhouse ovaries are cross-sectioned and the contents of all of the follicles are collected; mature oöcytes are collected, evaluated for quality, and used for fertilization. Immature oöcytes must be allowed to continue to develop in a maturation medium.

Either fresh or frozen-thawed semen can be used for fertilization. Sperm need to be capacitated in vitro in order to penetrate the zona pellucida and fuse with the ovum or to undergo the same maturation process that they would normally undergo in the female reproductive tract. Capacitation involves a series of cellular changes to the sperm including increased motility, calcium uptake and protein binding (binding to proteins produced by the female reproductive tract). In vitro capacitation is accomplished by creating a medium designed to simulate the female reproductive tract and allowing the sperm to incubate in it for a period of time. Sperm are then added to ova, incubated in culture medium for approximately 8-22 hours, and the resulting fertilized ova, called zygotes, are washed, examined for appropriate development, and allowed to continue to divide for up to seven days, again in culture. At that time, if embryos appear normal, they may either be frozen for future use or inserted into the uterus of a reproductively competent female.

The IETS reported that the number of in vitro embryos transferred in 2005 increased by 11 percent over 2004 numbers to more than 265,000. Almost all of these transfers were performed in South America (particularly in Brazil) and Asia (especially Korea and China).

The numbers of in vivo and in vitro embryo transfers for other species (sheep and goats) were considerably lower, with 34,458 sheep embryos transferred, mostly in Australia, New Zealand and South Africa, and approximately 5,000 goat embryos transferred, mostly in South Africa and Asia. According to IETS statistics, approximately 66,000 swine embryos were transferred in 2005 (a huge increase over the previous year's number of 16,016), with almost all occurring in Vietnam and Korea (Thibier 2006).

Embryo splitting

Genetically identical individuals derived from a sole embryonic source can arise naturally, as in the case of spontaneous monozygotic twinning, or in vitro via the manual separation (splitting) of early stage embryos. Embryo splitting may be considered the first true "cloning" procedure involving human intervention, and was first described by Willadsen and Polge in 1981, when monozygotic twin calves were produced.

Embryo splitting, or the mechanical separation of cells, can be used in very early embryos. Briefly, two-cell embryos derived from either in vitro fertilization, or embryo rescue following in vivo fertilization (as described for embryo transfer) are held in place with micropipettes under a microscope. The zona pellucida (the clear layer of protein surrounding the oöcyte and fertilized ovum) of these embryos is opened, and the two-celled embryo is then split into individual cells with a finely drawn needle or pipette. One of the cells is left in the original zona pellucida and the other is either placed into an empty zona pellucida or allowed to develop without a zona pellucida. These so-called demi-embryos can be cultured in vitro for a few days, inspected for appropriate growth and then transferred directly to synchronized recipient dams or frozen for future use. Similar procedures can be used to multiply embryos that have developed beyond the 2-cell stage (Willadsen 1980). Each of the individual cells or blastomeres from a single early embryo is totipotent. That is, each cell retains the ability to generate a fully functional individual identical to the other individual(s) derived from the other cells of the original embryo.

Although commercial applications of embryo splitting have been tracked by breeders' associations, the technology has never gained significant market penetration for several reasons. It is a very expensive and time consuming procedure that has not provided the yield initially anticipated for the technology. For example, actual calf yield from blastomere splits is approximately 105 calves per 100 embryos, while direct transfer of intact embryos yields approximately 60 calves per 100 embryos (Wilmut et al. 2000). Unless embryos are sexed at the time of splitting, however, breeders may end up with half of their animals being of the undesired gender, thus incurring twice the cost for the desired offspring. In addition, even if the resulting calves are of the desired gender, their production potential is not known, making the procedure an expensive gamble.

BLASTOMERE NUCLEAR TRANSFER

The next evolution of ART evolved from additional manipulations of the blastomere cell, and involved its fusion with an enucleated oöcyte. This method expands on the relatively simple early stage embryo cell separation procedure described previously by allowing the use of cells from later stage embryos. In this case, embryos of the eight to sixteen-cell stage, compact morulae, and the inner cell mass from blastocysts can be used as donor nuclei (First and Prather 1991a). Fusion of these later stage blastomere cells, which have lost their totipotency, with enucleated oöcytes, reprograms the blastomere nuclei to allow them to develop as zygotes. Blastomeres from bovine embryos up to the 64-cell stage can be fused with enucleated freshly fertilized oöcytes and cultured to develop into genetically identical individuals (Keefer et al. 1994). Cell nuclei derived from the inner cell mass of expanded blastocysts transferred into enucleated host cells are also capable of development resulting in offspring (Sims and First 1994).

This technology, which may be considered the true antecedent of somatic cell nuclear transfer, had limited commercial applicability for the same reasons as embryo splitting: high cost, high loss rate, and the inability to predict phenotypic performance or the gender of the resulting offspring.

Somatic Cell Nuclear Transfer (SCNT)

In 1962, biologist John Gurdon of Oxford University pioneered the method of the two step "nuclear transfer" process in frogs: the enucleation of a recipient oöcyte and the subsequent transfer of a differentiated somatic cell nucleus to that oöcyte. Gurdon's experiments showed that despite the differentiated status of the donor nucleus, reconstituted cells appeared to reprogram, or dedifferentiate, the nucleus and enable it to function much as a naturally produced zygote. These zygotes successfully developed into viable embryos that hatched and grew into tadpoles. Because the tadpoles had all come from the gut cells of the same adult frog, they all had the same genetic material and thus were all clones. However, Gurdon's nuclear transfer tadpoles clones failed to metamorphose into frogs. When scientists attempted to apply this technology to other species such as mice, cattle, or other mammals, the developmental program could not be reset (Gurdon and Uehlinger 1966; Byrne et al. 2002).

Scientists continued to tackle the problem and in 1986, Randall Prather and colleagues, then working in Neal First's laboratory at the University of Wisconsin-Madison, cloned a cow from early embryonic cells using nuclear transfer (Prather et al. 1987). Although this was an example of blastomere nuclear transfer, it effectively set the stage for Dolly's birth a decade later, on July 5, 1996. Dolly the sheep, the first organism ever to be cloned from adult cells, was created by Ian Wilmut and Keith Campbell using a technique similar to that used to create the first sheep from differentiated embryo cells (i.e., a blastomere clone) in 1995 (Wilmut et al. 1997).[16]

In July 1998, Ryuzo Yanagimachi, Toni Perry, and Teruhiko Wakayama of the University of Hawaii announced that they had cloned fifty mice from adult cells using the "Honolulu technique" (Wakayama et al. 1998). This was particularly significant because mouse embryos begin to divide almost immediately after the ovum is fertilized, and scientists had believed that this would not allow sufficient time for reprogramming to occur. Sheep, on the other hand, because their ova do not divide for several hours after fertilization, were thought to be an "easier" species to clone, as the natural delay between fertilization and division might be replicated in SCNT, possibly giving the oöcyte time to reprogram its new nucleus.

SCNT is a relatively new technology described by many as complex, technically demanding and inefficient, that continues to be developed and improved. As such, there is no set "method" that is universally employed, although the basic steps outlined below are common to most SCNT procedures at the time that this overview was written.

For species in which the cloning process has been relatively well developed, the first step is identifying the animal to use as a nuclear donor. Animals to be used for breeding purposes are selected because they have been shown to be genetically superior to herd mates for the trait(s) to be propagated. Somatic cells can be collected from the ear (hole punch) or skin (surgical incision or needle aspiration), although many other cell sources have been used. Multiple factors may influence success or failure of the nuclear transfer process. Coordination of the cell-cycle stage of the donor nucleus and the recipient egg cytoplasm appears to be important for successful development of embryos. In general, the selection of a cell type for commercial cloning from an adult animal has evolved to choosing a collection method that is relatively noninvasive and minimizes stress to the live animal donor.

Several characteristics have been identified as contributors to the degree to which any given donor cell or type of cell will likely result in a successful cloning event (Sung et al. 2006; Oback and Wells 2007a). One example is the "replicative state" of the donor cell. In general, cells in culture accumulate nutrients, grow, and when they reach certain conditions, divide. Cells that adhere to a solid substrate, such as the bottom of a tissue culture dish tend to grow until there are so many of them that they begin to touch each other. Once that happens, they generally stop dividing, and go into a "resting state" with respect to replication (referred to as G0). Cells can also be directed into G0 by depleting the nutrients in their growth medium. Some laboratories have concluded that cells in G0 are the most effective donors (Wilmut and Campbell 1998, De Sousa et al. 2002). Conversely, other laboratories have found that actively dividing cells make good donors (Cibelli et al. 1998a, Lanza et al. 2001). Some laboratories find that cells from embryos or fetuses are the best donors (Batchelder 2005), while others are successful at cloning cells from aged or even deceased animals (Hill et al. 2000a, Tian et al. 2001). Another characteristic that has been shown to influence the degree to which cells make good donors is how "inbred" the donor animal is (Rideout et al. 2000). These researchers have determined that "hybrid vigor" is important for the success rate of animal cloning and the more inbred the donor animal, the less likely it is that cloning will occur successfully. Further, some species appear to be more amenable to cloning than others (e.g., goats compared with cattle, see Chapter V), and some species have not been cloned at all. At this time, the best conclusion that can be drawn with respect to the degree to which a cell (or animal) will serve as a "good" donor is that the technology is not sufficiently mature to predict with certainty which set of conditions will optimize cloning efficiency.

OUTCOMES OBSERVED IN ARTS

As this risk assessment is being prepared, biologists are just beginning to understand the highly complex interactions that must occur to choreograph the millions of molecular interactions that signal the expression or silencing of genes in a particular cell or at any point in its life cycle. Although the exact mechanisms by which these effects occur are not fully understood, in all forms of reproduction, ranging from natural mating to SCNT, these processes may go awry in early development. Although most of the animals born following ARTs with significant in vitro components appear to be completely "normal," some of the outcomes are not so successful. In particular, some of the adverse outcomes noted in these "high in vitro component ARTs" appear to have common defects in gene expression, particularly in the overgrowth outcomes (Humpherys et al. 2002).

Published studies involving cattle, sheep, and mice demonstrate that embryos produced using in vitro systems may differ in morphology and developmental potential compared to embryos produced in vivo (Kruip and den Daas 1997; Young et al. 1998; Farin et al. 2004; Farin et al. 2006; see Appendix C for a more detailed discussion and additional references). For example, common abnormalities have been noted in fetuses (Farin and Farin 1995) and calves (Behboodi et al.1995; Sinclair 1999) associated with the transfer of bovine embryos produced using in vitro maturation (IVM), in vitro fertilization (IVF), in vitro culture (IVC) systems, and SCNT (Hill et al. 2000b). One set of reported adverse outcomes following transfer of embryos from cloning or in vitro production systems is often referred to as Large Offspring Syndrome (LOS). These include lowered pregnancy rates, increased rates of abortion, production of oversized calves, musculoskeletal deformities and disproportionalities, as well as hydroallantois (abnormal accumulation of fluid in the placenta) and other abnormalities of placental development.

The phenomenon of "large calves" was first described by Willadsen et al. (1991). The syndrome has also been identified in fetal and newborn lambs and in mice where the embryos were cultured in vitro (Eggan et al. 2001). Offspring with LOS tend to exhibit difficulties with placentation (Farin et al. 2004; Bertolini et al. 2004; Lee et al. 2004; Batchelder 2005). In cattle and sheep, the placentae of developing fetuses with LOS are unusually large for their species, and tend to have abnormal development of placentomes (the sites of attachment between fetal and maternally derived tissues of the placenta). LOS fetuses tend to have longer than usual gestation lengths, and often labor in the dams must be induced followed by Caesarian section deliveries. The newborns tend to be large for their breeds, and often have abnormal or poorly developed lungs, hearts, or other affected internal organs (liver and kidney), which makes it difficult for them to breathe or maintain normal circulation and metabolism. LOS newborns may appear to be edematous (fluid filled), and if they are to survive, often require significant veterinary intervention. Problems have also been noted in muscle and skeletal development of animals with LOS. These animals also often have difficulty regulating body temperature

Although the cause of LOS is not known with certainty, it is likely be related to changes in gene expression (i.e., epigenetic changes) that result from the in vitro manipulation and culturing of embryos. A review by Young et al. (1998) suggests that in vitro culture alone is adequate to perturb the embryo. This hypothesis is supported by data from Sinclair et al. (2000) where in vivo matured and fertilized eggs recovered from superovulated sheep donors, cultured in vitro for 6 days, showed an 18-36 percent increase in mean birth weight at day 125 of gestation, depending on the culture system used (Sinclair et al. 2000, Young et al. 1998).

This is an area of extensive research in the cloning and developmental biology communities. It is likely that advances in the understanding of these mechanisms will lead to significant improvements in the rates of successful outcomes of all ARTs that include a significant in vitro component, including cloning.

E. Future of Reproductive Technologies in Modern Agricultural Practice

Modern agricultural practices will likely continue to employ all of the reproductive modalities described in this overview. The factors that may influence which practices are used will likely be a function of the breeder/farmer/ranchers' needs and opportunities. Seidel (2006), in a foreword to a symposium on ARTs, emphasized that current differences in the reproductive management of cattle in different parts of the world are driven by multiple considerations. Some of these have to do with the nature of the differences in the husbandry of beef (mostly pasture based) and dairy (mostly intensively housed in the US, more pasture based in Australia and New Zealand) cattle. Some are economic (ARTs are much more expensive than natural matings), some are practical (the fertility of dairy cattle has declined significantly in the last 20 years, making ARTs more attractive; using ARTs in beef cattle is not practical for ranchers who look for replacement by natural coverage), and some are technological (the ability to choose genetics more precisely versus the developmental problems associated with ARTs with significant in vitro components).

Technological issues will be addressed by continued research and development in this field. To that end, several professional and scientific societies (e.g., the International Embryo Transfer Society, various animal science organizations and breeding associations) have been actively involved as clearing-houses for information and interaction.

SCNT has the potential to impact animal breeding in as fundamental a manner as artificial insemination. Given its current high costs (approximately $20,000 for a live calf) and relatively low success rates (< 10 percent), SCNT will likely be used to improve production characteristics of food producing animals by providing breeding animals, just as any breeding program would select the most elite animals for breeding, and not as production animals. In this way, cloning does not differ from any of the other ARTs that have been described in this chapter. Cloning has the relative advantage of allowing for the propagation of animals with known phenotypes to serve as additional breeding animals. This is critically important in breeding programs, especially when it may take years to "prove" the merit of a sire or dam. Second, it allows the propagation of animals whose reproductive function may be impaired. It has already been used to increase the available genotype of a particular dairy cow with low fertility; her clones appear to be exhibiting normal fertility (PIFB 2003). Third, it allows the propagation of valuable deceased animals from which tissue samples have been appropriately collected or preserved, which may have profound implications for species or breeds nearing extinction. Finally, for the first time, cloning allows for the careful study of the "nature-nurture" interactions that influence breeding programs by allowing a large enough sample of genetically identical animals to be raised in different environments, or with different diets. Such studies have been impossible to perform prior to the advent of SCNT and are likely to yield important information for developing livestock species to live in areas that have, until this time, been marginal for food animal production. This is of particular importance to the developing world, where even slightly increased wealth generally favors the incorporation of animal-based agriculture.

Regardless of the degree to which cloning may be adopted in animal breeding programs, FDA's role in performing this risk assessment is clear: the agency's responsibility is to determine whether cloning poses any risk to animals involved in the cloning process, and whether the consumption of food products from clones or their progeny poses any additional risk compared with food from conventionally produced animals. This Risk Assessment presents the method by which we evaluated data on clones and their progeny, the data themselves, and the agency's conclusions, including discussions of uncertainty.

REPRODUCTIVE TECHNOLOGY IN FARM ANIMALS

Farm animal selection and reproduction are on the threshold of the application of new biotechnologies. Modern biotechnologies will allow advances to be made. Research into physiology and embryology has provided a basis for the development of technologies that increase productivity of farm animals through enhanced control of reproductive function. The livestock provides many opportunities to utilize these disciplines and evolving competencies. Artificial insemination, embryo transfer, in vitro fertilization, cloning, transgenics and genomics all are components of the tool box for present and future applications. Understanding the mechanisms that regulate reproductive function has important implications for this diverse field. Several peptides play a role in determining the normal functioning of the neuroendocrine regulation of reproduction. Kiss1 neurons have emerged as primary transducers of internal and environmental cues to regulate the neuroendocrine reproductive axis. Leptin serves as a metabolic signal that acts on the hypothalamic-pituitary-ovarian axis to enhance GnRH and LH secretion and ovarian function. Leptin effects on Gonadotropin-Releasing Hormone (GnRH) /LH secretion are mediated by NPY and kisspeptin. In recent years, livestock productivity has been increased by improved reproduction. Various techniques have been developed and refined to obtain a large number of offspring from genetically superior animals or obtain offspring from infertile animals. These techniques include: artificial insemination, cryopreservation of gametes or embryos, induction of multiple ovulations, embryo transfer, in vitro fertilization, sex determination of sperm or embryos, nuclear transfer, cloning, etc. Further the wide development radio-immuno-assay technology offers wide scope for improving the reproductive efficiency of farm animals. RIA technique for early non-pregnancy diagnosis can be integrated in to AI programmes in order to increase their effectiveness, reduce the unproductive period of dairy cows and increase the economic benefits to farmers. The greater challenge lies ahead for animal researchers is to integrate and potentially exploit these novel technologies in a society-friendly manner. Accepting this challenge and working towards achieving such targets should enable us to reap maximum benefits from the farm animal sector.

INTRODUCTION

Remarkable improvements in livestock productivity in the developed countries have achieved through research into how animals grow and how yields can be influenced and protected, followed by the widespread uptake of new techniques and materials. Many of the advances in improving the feeding, fertility and health of livestock have been possible with use of nuclear techniques (Naqvi et al., 2002; Khanal and Munankarmy, 2009). Farm animal selection and reproduction are on the threshold of the application of new biotechnologies. Modern biotechnologies will allow advances to be made. Globally, there is active competition for new technologies that may directly or indirectly affect the future of animal production, including breeding. The combined advances in genetics, embryology and stem cell research open the prospect that high-added value products for agricultural, medical and technical applications will soon become a reality (Srivastava and Sejian, 2010; Meenambigai and Sejian, 2011).

Reproductive technology encompasses all current and anticipated uses of technology in human and animal reproduction, including assisted reproductive technology, contraception and others (Mapletoft and Hasler, 2005). Research into physiology and embryology has provided a basis for the development of technologies that increase productivity of farm animals through enhanced control of reproductive function. Animal Biotechnology represents an expanding collection of rapidly developing disciplines in science and information technologies. The livestock provides many opportunities to utilize these disciplines and evolving competencies. Artificial insemination, embryo transfer, in vitrofertilization, cloning, transgenics and genomics all are components of the tool box for present and future applications (Betteridge, 2003). Individually, these are powerful tools capable of providing significant improvements in productivity. Combinations of these technologies coupled with information systems and data analysis will provide even more significant changes in the next decade. Various techniques have been developed and refined to obtain a large number of offspring from genetically superior animals or obtain offspring from infertile (or sub fertile) animals (Naqvi et al., 2001; Blackburn, 2004). To take full advantage of the benefits of assisted reproductive technologies, one must understand the basic physiology of the female and male reproductive systems as well as various methods to synchronize reproductive cycles (Paterson et al., 2003).

Based on the progress in scientific knowledge of endocrinology, reproductive physiology, cell biology and embryology during the last fifty years new biotechniques have been developed for and introduced into animal breeding and husbandry (Wrathall et al., 2004). Among them are oestrus synchronization/ induction, artificial insemination, Multiple Ovulation Induction and Embryo Transfer (MOET), in vitro embryo production (IVP) and cloning by Nuclear Transfer (NT). The aims of these reproductive technologies were initially to speed up the genetic improvements of farm animals by the increase of offspring of selected males and females and the reduction of the generation intervals. The technique of cloning by nuclear transfer is mainly applied for experimental purposes, with the prospect of a more practical implementation in the near future, with the aims of the enhancement of the uniformity of herds for an easier management or for the multiplication of transgenic animals after gene-targeting (Polejaeva and Campbell, 2000; Meenambigai et al., 2009).

Techniques are now available to get genetically improved farm animals in large numbers. This involves collection of the fertilized egg from genetically superior female which ovulate spontaneously or are induced to superovulate (Wolf et al., 2000). The above goal can also be achieved by transferring the embryo of genetically improved variety into genetically less desirable females, thus getting many more progeny of superior individuals. The current efficiency for producing transgenic animals particularly farm animals, is low and the cost is high. Success in the production of transgenic farm animals requires an adequate animal facility and dedicated teams of embryologists, veterinarians, animal scientists and molecular biologists. Improvement on the success rate of the transgenic technology largely relies on the application of (1) various classical reproductive and embryological technologies such as artificial insemination, superovulation and reproductive management etc. and (2) newly emerged contemporal technologies such cloning and other assisted reproductive technologies.

NEUROENDOCRINE REGULATION OF REPRODUCTION

Understanding the mechanisms that regulate reproductive function has important implications for diverse field. The neuroendocrine pathway regulating reproductive status is the Hypothalamo-Pituitary-Gonadal (HPG) axis. Reproductive activity is induced by the release of hypothalamic gonadotropin-releasing hormone (GnRH) and the subsequent release of pituitary gonadotropins, Luteinizing Hormone (LH) and Follicle-Stimulating Hormone (FSH). LH and FSH act on the gonads to induce the production of sex steroids and gamete development. Sex steroids, in turn, act on the brain to promote appropriate sexual behaviour. describes the neuroendocrine regulation of reproduction in male and female animals. Excitatory amino acids (ExAA) such as glutamate and aspartate are important neurotransmitters that play an important role in neuroendocrine control of anterior pituitary hormone secretion. The gonadotropins LH and FSH control testicular function, including testosterone secretion and spermatogenesis. In male domestic animal species, ExAA agonists, such as n-methyl-D, L-aspartate (NMA), effectively stimulate LH secretion and subsequent testosterone release, by acting within the brain to stimulate secretion of GnRH. Stimulatory effects of ExAA agonists on FSH secretion, however, have not been demonstrated.

Inhibins and oestradiol act directly at the anterior pituitary to decrease expression of the gene encoding the FSH subunit. They act to reduce both transcription and stability of mRNA, effects which override GnRH action on FSH release. Oestradiol causes a major decrease in FSH, while LH is initially decreased and then increased.

The initial stages of folliculogenesis occur independently of gonadotrophic hormones. GnRH binds to specific receptors on gonadotrophs and triggers the release of intracellular Ca^{2+}, causing a transient release of both FSH and LH. FSH is the key hormone stimulating the emergence of waves of follicles and its decline is associated with selection of a dominant follicle, which then becomes dependent on LH for its final fate when concentrations of FSH are minimal (Roche, 1996). During the oestrous cycle of ewes, the patterns of secretion of LH and FSH are different, with an increase in LH pulse frequency of 1 per hour during pro-oestrus, while there is a coincident decrease in FSH due to the increase in oestradiol and perhaps inhibins. There is a major depletion of LH after the preovulatory surge. Thus, LH release is a regulated pathway mediated by GnRH action, whereas FSH release is a more constitutive pathway, in which synthesis is followed by release rather than by storage.

Neurons that produce gonadotropin-releasing hormone(GnRH) reside in the basal forebrain and drive reproductive function in mammals. Kisspeptin, the peptide encoded by the KiSS-1 gene, plays an important role in the development and upregulation of the reproductive system (Funes et al., 2003; Seminara et al., 2003; Greives et al., 2008), whereas GnIH has been reported to play an important role in regulating reproduction by downregulating release of pituitary gonadotropins via actions on the HPG axis. Kisspeptin neurons involved in control of reproduction are situated in anteroventral periventricular (AVPV) and arcuate (ARC) nuclei of the brain. The receptor for kisspeptin, G-proteincoupled receptor 54 (GPR54), has been localized to a majority of GnRH neurons (Quaynor et al., 2007). It has been established that kisspeptin produces its effects via actions on GnRH neurons (Smith and Clarke, 2007).

FARM ANIMAL CLONING

General description of cloning technology: Clone is an organism or animal that is having the genetic material identical to the ancestor from which its genome has descended. The term cloning is the technique applied to produce identical organisms or animals that have similar genetic material. It can be achieved by two ways. First is by splitting the embryo at an early (pluripotent) stage so that identical twins will be produced. This technique has been performed first in sea urchins, followed by salamanders and farm animals. Initially embryo splitting gained much of importance but later lost the track as it is technically complicated and less efficient because one ovum can give rise to only two identical individuals. Second method is by somatic cell nuclear transfer. The concept of cloning an animal by transfer of somatic cell nucleus was proposed in 1938. Slowly this has been tested and proved to be feasible first in amphibians followed by the demonstration in mice. This technique gained much focus in 1997 when, cloned the first sheep Dolly portraying this as an efficient method for farm animal production. In this technique the somatic cell nucleus is transferred into the enucleated cytoplasm of an oocyte. After the transfer of nucleus to the enucleated ovum, the ovum is transferred to the womb of a surrogate mother to activate the embryo to start dividing.

Pros and cons of cloning technology: Animal cloning ensures the sustainability of the desirable phenotype allowing them for better production of high quality and safe food products. At present the main reason to clone farm animals is to preserve the breeding capacity of genetically elite animals (proven through progeny testing), particularly males and to insure against loss of valuable genetic and characteristic features. Pig cloning involves the use of valuable boars for two reasons. First, because billions of sperm are required for artificial insemination and so any given boar can serve only a small number of females compared with bulls. Second, the most effective way to evaluate the genetic quality of a pig is through a detailed analysis of the carcass. Cloning offers the opportunity to evaluate the quality of a boar by slaughtering it and then using numerous copies of those individuals whose carcasses meet the required standards. In females similar reasons can be given and so oocytes can also be valuable. For other animals it may be to restore their breeding capacity e.g. after injury, disease, old age, or that they have been previously castrated (horses, dogs, cats).

Employing this technique, animals mimicking human diseases can be cloned and studied in a better way which would be directly advantageous for human welfare. Cloning is being increasingly used in developmental biology; in the aetiology and pathophysiology of disease; in biotechnology for the production of therapeutic agents e.g., pharmaceutical proteins, organ development, in xenotransplantation and possibly in safety testing.

Cloning can bring a revolution in the meat industry as the necessity and cost involved in rearing animals is cut off. Cloning could be useful in preserving the endangered species which are in the verge of extinction. The enucleated oocytes of an animal were used to produce a cloned embryo. In the case of extinct breeds cryopreserved DNA, naturally occurring samples or preserved tissues could possibly be used providing an appropriate surrogate dam can be found. The success rate of cloning is very less due to the decreased efficiency. 90% of nuclear transfers turn out to be unproductive. Added to this, cloned animals tend to have more compromised immune function and higher rates of infection, tumor growth and other disorders. There are reports that cloned mice live in poor health and die early. Studies have shown that about 4% of genes function abnormally. The abnormalities do not arise from mutations in the genes but from changes in the normal activation or expression of certain genes.

Also, it is not clear whether the nuclear reprogramming that takes place during the cloning process has any influence over the animal food products that arise from cloned farm animals.

Problems frequently encountered with cloning: Placental abnormalities, fetal overgrowth, prolonged gestation, stillbirth, hypoxia, respiratory failure and circulatory problems, lack of post-natal vigour, increased body temperature at birth, malformations in the urogenital tract (hydronephrosis, testicular hypoplasia), malformations in liver and brain, immune dysfunction, lymphoid hypoplasia, anaemia, thymic atrophy and bacterial and viral infection.

NUCLEAR TECHNIQUES IN ANIMAL REPRODUCTION

In recent years, livestock productivity has been increased by improved reproduction. Various techniques have been developed and refined to obtain a large number of offspring from genetically superior animals or obtain offspring from infertile animals. These techniques include: artificial insemination, cryopreservation of gametes or embryos, induction of multiple ovulations, embryo transfer, fertilization, sex determination of sperm or embryos, nuclear transfer, cloning, etc.

Nuclear transfer: Nuclear Transfer (NT), has numerous potential applications and considerable impact, mainly in agriculture, medicine, pharmacy and fundamental biology. In addition, somatic cell nuclear transfer is the most efficient alternative to produce large transgenic animals. The potential applications of Somatic Cell Nuclear Transfer (SCNT) technique have received much attention since the first births of cloned animals were reported in various domestic species, including sheep , cattle , goat , pig and more recently, horse . With the nuclear transfer technology one can now create some domestic animals with specific genetic modifications. An ever-expanding variety of cell types have been successfully used as donors to create the clones. Both cell fusion and microinjection are successfully being used to create these animals. However, it is still not clear which stage(s) of the cell cycle for donor and recipient cells yield the greatest degree of development .

In agriculture, there is current interest in cloning selected livestock that are rare or that have desired genetic traits, such as disease resistance, improved meat quality or yield, or increased milk production. These cloned livestock would be used in breeding programs to incorporate their positive genetic traits into herds more quickly than is possible through conventional breeding and artificial insemination. Developmental biologists in agrobiotechnology have been trying for years to find conditions under which nuclear material from somatic diploid cells could be used as pluripotent source of genetic information. If the donor cell could be genetically manipulated in cell culture using standard stable DNA transfection protocols, then transgenic farm animals would be a common as transgenic mice .Since, the first mammals to be cloned from cultured differentiated cells and the birth of Dolly, the first mammal derived from an adult somatic cell progress was fast.Nuclear replacement efficiency varies among species.

Somatic cell nuclear transfer is characterized by a series of developmental abnormalities, the so called cloning-syndrome. The incidence of these anomalies varies according to the species, genotype, donor cell status, or specific aspects of the nuclear transfer and culture protocols used. Despite the present inefficiencies, mouse nuclear replacement experiments have proven that this method is already capable of improving the efficiency of transgenic production compared to traditional methods. Nuclear transfer is a fast developing technique, far from being optimized. Evolution in the efficiency and applicability in new species is expected. However, it is not clear how much of an improvement can be achieved by simply optimizing the present protocols. opined that revolutionary changes in our understanding of the reprogramming of the nuclear material, organization of chromatin and epigenetic changes are necessary to allow proper control of the procedures and to avoid epigenetic aberrations in the progeny.

The potential benefits of nuclear transfer are: (1) All animals born will be transgenic, (2) Faster development of a production flock or herd, significantly reducing the time to production of protein or peptide. Where it would normally require 44 months to reach production flock status in sheep, (78 months in cows), nuclear transfer technology can reach production flock status in 18 months for sheep, (33 months for cows); (3) Earlier availability of product, accelerating progress to clinical trials; (4) The selective removal of gene(s), replacing them with the desired human equivalents; (5) Increased expression levels of product (by targeting inserted genes to sites of high expression); (6) Cellular-level analysis of clones allowing pre-selection for optimal protein expression.

Somatic cell cloning by nuclear transfer is a relatively new technology with many potential applications. However, at the current stage of development, the reprogramming of epigenetic inheritance by nuclear transfer is still incomplete. Further efforts and new paradigms are needed to perfect this technology and extend it to its fullest potential. Various strategies have been used to improve the efficiency of nuclear transfer, however, significant breakthroughs are yet to happen.

Gene transfer: Gene transfer is to transfer a gene from one DNA molecule to another DNA molecule. There are two basic approaches presently in use for inserting DNA into vertebrate germ line cells-transfection and infection with retrovirus vectors. A third approach based on the use of mobile genetic elements, has been commonly used for insects and is being explored for germ-line modification of vertebrates.

Some of the more promising alternative strategies such as sperm-mediated gene transfer, restriction enzyme-mediated integration, metaphase II transgenesis and a new twist on retrovirus-mediated gene transfer have also been recently identified. To be functional, the integrated gene must be expressed and regulated appropriately. Thus, the gene to be transferred must be accompanied by the appropriate promoter and regulatory sequences. Some genes require an enhancer that may be located far from the promoter. Gene that is introduced into the cell from an external source is called transgene. A fertile animal that carries an introduced gene(s) in its germ line is called transgenic animal. The introduction of gene(s) into animal cells that leads to the transmission of the input transgene to successive generation is called transgenesis.

Marker-assisted selection: The potential benefits of using markers linked to genes of interest in breeding programmes, thus moving from phenotype based towards genotype-based selection, have been obvious for many decades. However, realization of this potential has been limited by the lack of markers. With the advent of DNA-based genetic markers in the late 1970s, the situation changed and researchers could, for the first time, begin to identify large numbers of markers dispersed throughout the genetic material of any species of interest and use the markers to detect associations with traits of interest, thus allowing MAS finally to become a reality.'

Molecular marker maps, the necessary framework for any MAS programme, have been constructed for the majority of agriculturally important species, although the density of the maps varies considerably among species. Despite the considerable resources that have been invested in this field and despite the enormous potential it still represents, with few exceptions, MAS has not yet delivered its expected benefits in commercial breeding programmes for livestock in the developed world. When evaluating the potential merits of applying MAS as a tool for genetic improvement in developing countries, some of the issues that should be considered are its economic costs and benefits, its potential benefits compared with conventional breeding or with application of other biotechnologies and the potential impact of Intellectual Property Rights (IPRs) on the development and application of MAS.

Sex determination of sperm: There has been great interest in sexing sperm ever since AI was practiced widely. The basic principles are simple; the X-sperm contains more DNA than the Y-sperm. Although this difference is small, by attention to details, it is possible to measure DNA content of individual sperm with sufficient accuracy to distinguish between X- and Y-sperm. A number of approaches to the sexing of semen have been attempted and several have been reported as successful. However, the only method of semen sexing that has shown any promise has been the sorting of spermatozoa according to the DNA content, by means of flow cytometry. Flow cytometric technology is widely used for sexing sperm in mammalian species. Differences in DNA content have provided both a method to differentiate between these sex-determining gametes and a method to sort them that can be used for predetermining sex in mammals. The precision of this DNA measurement is such that the difference in DNA content between mammalian X-and Y-chromosome bearing sperm of a variety of mammals has been determined. Successful sperm sexing of various species must take into account the relative susceptibilities of gametes to the stresses that occur during sexing. Sperm sexing by high speed flow cytometry has been one of the most significant new technologies for artificial breeding of livestock developed in the twentieth century . A very recent and most exciting development in this field is the establishment of reverse sex-sorting technology for the utilization of frozen ram and bull semen .These results demonstrate that frozen-thawed ram and bull sperm can be sex-sorted for either immediate or future use in an IVF system after re-cryopreservation.

Sex determination of embryos: Producers of domestic livestock strive to improve genetic influences in their herds. This requires identification and propagation of animals that demonstrate desirable characteristics. The more animals available from which to select, the greater the opportunity to discover high-performance animals. Predetermination of the sex of offspring would provide a greater number of males or females from which to select the top individuals that will contribute the genetics to the next generation. The embryo transfer technology represents a powerful tool for the acceleration of various breeding programs in cattle. Known sex of embryos produced for use in ET programs can more effectively help to manage producer resources because more heifer calves per ET can be produced. This approach can improve the genetic potential of cattle herds in shorter time intervals. Sexing of embryos before transfer and implanting has great potential for the livestock industry. Manipulating the sex of offspring has been a dream of livestock industry for decades.

Embryo sexing has been attempted by a variety of methods, including cytogenetic analysis, assays for X-linked enzyme activity, analysis of differential development rates, detection of male-specific antigens and the use of Y-specific DNA probes. In DNA amplification techniques, the probe itself is not used. The sequence information generated from the identification and cloning of the probe is used instead. Several methods have now been reported for mammalian sex determination as following (1) Using polymerase chain reaction (PCR) to amplify a Y-chromosome; (2) PCR-based genotyping; (3) PCR-mediated approach uses three sets of primers for sex determination of pre-implantation bovine embryos; (4) rapid sexing method for preimplantation embryos of bovine using Loop-Mediated Isothermal Amplification (LAMP) reaction.

Since, the development of DNA amplification techniques, particularly the Polymerase Chain Reaction (PCR), these techniques have been applied to numerous situations in which the analysis of rare sequences is desired. It was the identification of bovine Y-chromosome specific DNA probes and the subsequent development of DNA amplification techniques by PCR, that made the possibility of sexing embryos into a reality. Using DNA analysis to analyse the sex of embryos has been shown to be reliable. The removal of a few cells in this procedure, caused very little trauma to the embryos. It did not alter developmental potential in vitro. The procedure has been shown to be sensitive, accurate and efficient and pregnancy rates have not been affected, compared to those after transfer of fresh embryos without sexing. Several commercial Polymerase Chain Reaction (PCR) kits are available which uses the primers specific to the Y-Chromosome Determinant (YCD) to determine sex of the embryos. The PCR product was detected by UV light in agarose gel with ethidium bromide and the embryos were scored as Y-chromosome determinant positive (male) or Y-chromosome determinant negative.

Radioimmunoassay impact on animal reproduction: Early pregnancy detection and prediction of the number of fetus would be profitable for livestock breeders because it enables them to adjust nourishment of pregnant animals according to the individual needs in order to prevent health problems around parturition. RIA technique for early non-pregnancy diagnosis can be integrated in to AI programmes in order to increase their effectiveness, reduce the unproductive period of dairy cows and increase the economic benefits to farmers. ProgesteroneRIA presented a very reliable tool for early diagnosis of non-pregnancy and infertility. The use of progesterone measurement for early diagnosis of non-pregnant cows has a high level of accuracy and can improve the efficiency of AI delivery, reduce calving intervals and improve milk and meat production from the dairy industry. Further monitoring reproduction through RIA of progesterone was cost-effective in improving reproductive efficiency. In addition RIA technique can be used to detect physiologically relevant increases in serum FSH related to emergence of each new wave of follicle growth. Highly sensitive RIA using I-125 has been developed and used to measure the minute quantities of reproductive and other hormones circulating in the blood that control reproduction. Development of RIA technology has paved way for estimation of almost all reproductive hormones which in turn depicts the reproductive status of the animals. RIA has made it possible to determine when animals are ready for breeding, diagnose pregnancy earlier than would be otherwise possible, check whether animals have been inseminated at the correct time, devise corrective measures for reproductive disorders and improve the efficiency of artificial insemination and embryo transfer programs.

7

Elementary Anatomy and Physiology

Veterinary nurses need to have a firm grasp of the normal structure of an animal's body and how it functions before they can understand the effect diseases and injuries have and the best ways to treat them.

This chapter describes the structure of the animal body and the way in which it works. Animals encountered in normal veterinary practice are used as examples where possible. The chapter is divided into 16 topic:

- Chemicals
- Classification
- The Cell
- Body Organisation
- The Skin
- The Skeleton
- Muscles
- Cardiovascular System
- Respiratory System
- Lymphatic System
- The Gut and Digestion
- Urinary System
- Reproductive System
- Nervous System
- The Senses
- Endocrine System

CHEMICALS

After completing this section, you should know the:

- symbols used to represent atoms;
- names of molecules commonly found in animal cells;
- characteristics of ions and electrolytes;
- basic structure of carbohydrates with examples;
- carbohydrates can be divided into mono- di- and poly-saccharides;
- basic structure of fats or lipids with examples;
- basic structure of proteins with examples;
- function of carbohydrates, lipids and proteins in the cell and animals' bodies;
- foods which supply carbohydrates, lipids and proteins in animal diets.

Elements And Atoms

The elements (simplest chemical substances) found in an animal's body are all made of basic building blocks or atoms. The most common elements found in cells are given in the table below with the symbol that is used to represent them.

Atom	**Symbol**
Calcium	Ca
Carbon	C
Chlorine	Cl
Copper	Cu
Iodine	I
Hydrogen	H
Iron	Fe
Magnesium	Mg
Nitrogen	N
Atom	Symbol
Oxygen	O
Phosphorous	P
Potassium	K
Sodium	Na
Sulfur	S
Zinc	Zn

Compounds And Molecules

A compound or molecule is formed when two or more atoms join together. Note that some atoms are never found alone. For example oxygen is always found as molecules of 2 oxygen atoms (represented as O2).

The table below gives some common compounds.

Compound	Symbol
Calcium carbonate	$CaCO_3$
Carbon dioxide	CO_2
Copper sulfate	$CuSO_4$
Glucose	$C_6H_{12}O_6$
Hydrochloric acid	HCl
Sodium bicarbonate (baking soda)	$NaHCO_3$
Sodium chloride (table salt)	NaCl
Sodium hydroxide	NaOH
Water	H_2O

Chemical Reactions

Reactions occur when atoms combine with or separate from other atoms. In the process new products with different chemical properties are formed.

Chemical reactions can be represented by chemical equations. The starting atoms or compounds are usually put on the left-hand side of the equation and the products on the right-hand side.

For example

- $H_2O + CO_2$ gives H_2CO_3
- Water + Carbon dioxide gives Carbonic acid

Ionization

When some atoms dissolve in water they become charged particles called ions. Some become positively charged ions and others negatively charged. Ions may have one, two or sometimes three charges.

The table below shows examples of positively and negatively charged ions with the number of their charges.

	Positive Ions	Negative Ions	
H+	Hydrogen	Cl-	Chloride
Ca+	Calcium	OH-	Hydroxyl
Na+	Sodium	HCO_3-	Bicarbonate
K+	Potassium	CO_3-	Carbonate
Mg+	Magnesium	SO_4-	Sulfate
Fe+	Iron (ferrous)	PO_4-	Phosphate
Fe+	Iron (ferric)	S-	Sulfide

Positive and negative ions attract one another to hold compounds together.

Ions are important in cells because they conduct electricity when dissolved in water. Substances that ionise in this way are known as electrolytes.

The molecules in an animal's body fall into two groups: inorganic compounds and organic compounds. The difference between these is that the first type does not contain carbon and the second type does.

Organic And Inorganic Compounds

Inorganic compounds include water, sodium chloride, potassium hydroxide and calcium phosphate.

Water is the most abundant inorganic compound, making up over 60% of the volume of cells and over 90% of body fluids like blood. Many substances dissolve in water and all the chemical reactions that take place in the body do so when dissolved in water. Other inorganic molecules help keep the acid/base balance (pH) and concentration of the blood and other body fluids stable (8).

Organic compounds include carbohydrates, proteins and fats. All organic molecules contain carbon atoms and they tend to be larger and more complex molecules than inorganic ones. This is largely because each carbon atom can link with four other atoms. Organic compounds can therefore consist of from one to many thousands of carbon atoms joined to form chains, branched chains and rings. All organic compounds also contain hydrogen and they may also contain other elements.

A carbon atom with 4 links

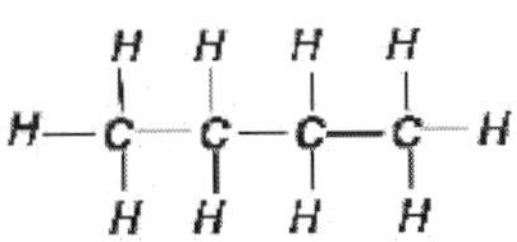

A chain of carbon atoms with attached hydrogens

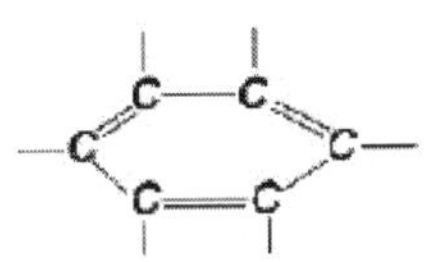

A ring of carbon atoms

Carbohydrates

The name "carbohydrate" tells you something about the composition of these "hydrated carbon" compounds. They contain carbon, hydrogen and oxygen and like water (H2O), there are always twice as many hydrogen atoms as oxygen atoms in each molecule. Carbohydrates are a large and diverse group that includes sugars, starches, glycogen and cellulose. Carbohydrates in the diet supply an animal with much of its energy and in the animal's body, they transport and store energy.

Carbohydrates are divided into three major groups based on size: monosaccharides (single sugars), disaccharides (double sugars) and polysaccharides (multi sugars).

Monosaccharides are the smallest carbohydrate molecules. The most important monosaccharide is glucose which supplies much of the energy in the cell. It consists of a ring of 6 carbon atoms with oxygen and hydrogen atoms attached.

Disaccharides are formed when 2 monosaccharides join together. Sucrose (table sugar), maltose, and lactose (milk sugar), are three important disaccharides. They are broken down to monosaccharides by digestive enzymes in the gut.

Polysaccharides like starch, glycogen and cellulose are formed by tens or hundreds of monosaccharides linking together. Unlike mono- and di-saccharides, polysaccharides are not sweet to taste and most do not dissolve in water.

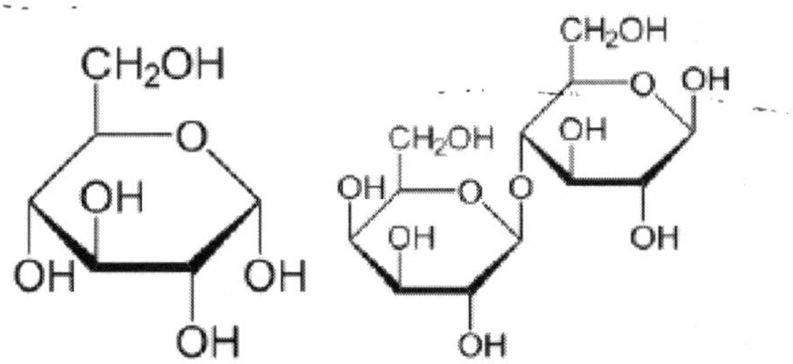

Disaccharide (Lactose)

- Starch is the main molecule in which plants store the energy gained from the sun. It is found in grains like barley and roots like potatoes.
- Glycogen, the polysaccharide used by animals to store energy, is found in the liver and the muscles that move the skeleton.
- Cellulose forms the rigid cell walls of plants. Its structure is similar to glycogen, but it can't be digested by mammals. Cows and horses can eat cellulose with the help of bacteria which live in specialised parts of their gut.

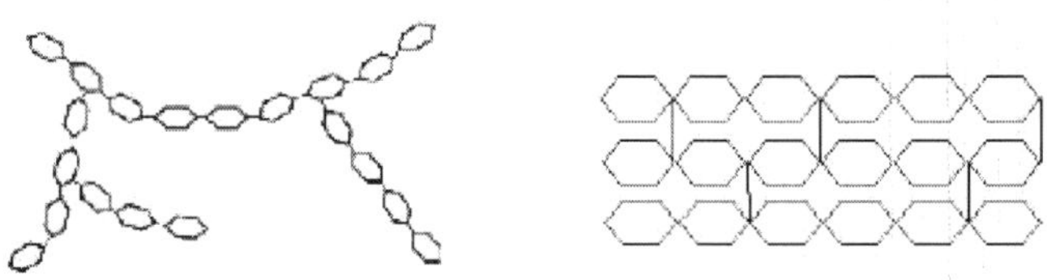

glycogen cellulose

(note that each ring is a monosaccharide)

Fats

Fats or lipids are important in the plasma membrane around cells and form the insulating fat layer under the skin. They are also a highly concentrated source of energy, and when eaten in the diet provide twice as much energy per gram as either carbohydrates or proteins.

Like carbohydrates fats contain carbon, hydrogen and oxygen, but unlike them, there is no particular relationship between the number of hydrogen and oxygen atoms.

The fats and oils animals eat in their diets are called triglycerides or neutral fats. The building blocks of triglycerides are 3 fatty acids attached to a backbone of glycerol (glycerine). When fats are eaten the digestive enzymes break down the molecules into separate fatty acids and glycerol again.

Fatty acids are divided into two kinds: saturated and unsaturated fatty acids depending on how much hydrogen they contain. The fat found in animals bodies and in dairy products contains mainly saturated fatty acids and tends to be solid at room temperature. Fish and poultry fats and plant oils contain mostly unsaturated fatty acids and are more liquid at room temperature.

Phospholipids are lipids that contain a phosphate group. They are important in the plasma membrane of the cell.

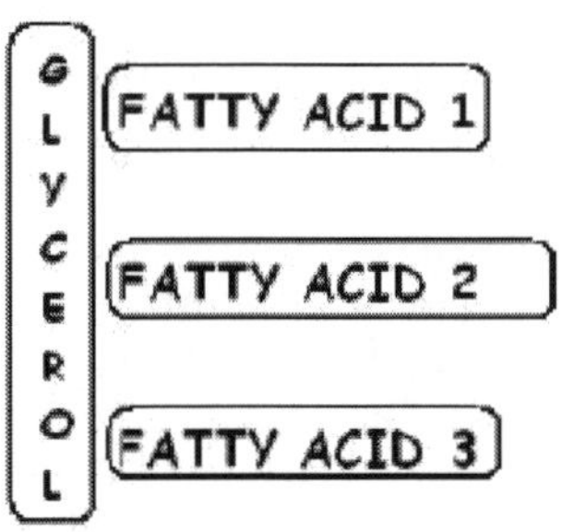

A chain of amino acids

Proteins

Proteins are the third main group of organic compounds in the cell - in fact if you dried out a cell you would find that about 2/3 of the dry dust you were left with would consist of protein. Like carbohydrates and fats, proteins contain C, H and O but they all also contain nitrogen. Many also contain sulphur and phosphorus.

In the cell, proteins are an important part of the plasma membrane of the cell, but their most essential role is as enzymes. These are molecules that act as biological catalysts and are necessary for biochemical reactions to proceed. Protein is also found as keratin in the skin, feathers and hair, in muscles, as well as in antibodies and some hormones.

Proteins are built up of long chains of smaller molecules called amino acids. There are 20 common types of amino acid and different numbers of these arranged in different orders create the multitude of individual proteins that exist in an animal's body.

Long chains of amino acids often link with other amino acid chains and the resulting protein molecule may twist, spiral and fold up to make a complex 3-dimensional shape. As an example, see the diagram of the protein lysozyme below. Believe it or not, this is a small and relatively simple protein.

t is this shape that determines how proteins behave in cells, particularly when they are acting as enzymes. If for any reason this shape is altered, the protein stops working. This is what happens if proteins are heated or put in a solution that is too acidic or alkaline. Think what happens to the "white'" of an egg when it is cooked. The "white" contains the protein albumin, which is changed or "denatured" permanently by cooking. The catastrophic effect that heat has on enzymes is one of the reasons animals die if exposed to high temperatures.

In the animal's diet, proteins are found in meat (muscle), dairy products, seeds, nuts and legumes like soya. When the enzymes in the gut digest proteins they break them down into the separate amino acids, which are small enough to be absorbed into the blood.

CLASSIFICATION

Classification is the process used by scientists to make sense of the 1.5 million or so different kinds of living organisms on the planet. It does this by describing, naming and placing them in different groups. As veterinary nurses you are mainly concerned with the Animal Kingdom but don't forget that animals rely on the Plant Kingdom for food to survive. Also many diseases that animals are affected by are members of the other Kingdoms -- fungi, bacteria and single celled animals.

Naming And Classifying Animals

There are more than 1.5 million different kinds of living organism on Earth ranging from small and simple bacteria to large, complex mammals. From the earliest time that humans have studied the natural world they have named these living organisms and placed them in different groups on the basis of their similarities and differences.

Naming Animals

Of course we know what a cat, a dog and a whale are but, in some situations using the common names for animals can be confusing. Problems arise because people in different countries, and even sometimes in the same country, have different common names for the same animals. For example a cat can be a chat, a Katze, gato, katt, or a moggie, depending on which language you use. To add to the confusion sometimes the same name is used for different animals. For example, the name 'gopher' is used for ground squirrels, rodents (pocket gophers), for moles and in the south-eastern United States for a turtle. This is the reason why all animals have been given an official scientific or binomial name. Unfortunately these names are always in Latin. For example:

- Common rat: Rattus rattus
- Human: Homo sapiens
- Domestic cat: Felis domesticus
- Domestic dog: Canis familiaris

As you can see from the above there are certain rules about writing scientific names:

- They always have 2 parts to them.
- The first part is the genus name and is always written with a capital first letter.
- The second name is the species name and is always written in lower case.
- The name is always underlined or printed in italics.

The first time you refer to an organism you should write the whole name in full. If you need to keep referring to the same organism you can then abbreviate the genus name to just the initial. Thus "Canis familiaris" becomes "C. familiaris" the second and subsequent times you refer to it.

Classification of Living Organisms

To make some sense of the multitude of living organisms they have been placed in different groups. The method that has been agreed by biologists for doing this is called the classification system. The system is based on the assumption that the process of evolution has, over the millennia, brought about slow changes that have converted simple one-celled organisms to complex multi-celled ones and generated the earth's incredible diversity of life forms. The classification system attempts to reflect the evolutionary relationships between organisms.

Initially this classification was based only on the appearance of the organism. However, the development of new techniques has advanced our scientific knowledge. The light microscope and later the electron microscope have enabled us to view the smallest structures, and now techniques for comparing DNA have begun to clarify still further the relationships between organisms. In the light of the advances in knowledge the classification has undergone numerous revisions over time.

At present most biologists divide the living world into 5 kingdoms, namely:

- bacteria
- protists
- fungi
- plants
- animal

We are concerned here almost entirely with the Animal Kingdom. However, we must not forget that bacteria, protists, and fungi cause many of the serious diseases that affect animals, and all animals rely either directly or indirectly on the plant world for their nourishment.

The Animal Kingdom

So what are animals? If we were suddenly confronted with an animal we had never seen in our lives before, how would we know it was not a plant or even a fungus? We all intuitively know part of the answer to this.

Animals

- eat organic material (plants or other animals)
- move to find food
- take the food into their bodies and then digest it
- and most reproduce by fertilizing eggs by sperm

If you were tempted to add that animals are furry, run around on four legs and give birth to young that they feed on milk you were thinking only of mammals and forgetting temporarily that frogs, snakes and crocodiles, birds as well as fish, are also animals.

These are all members of the group called the vertebrates (or animals with a backbone) and mammals make up only about 8% of this group. The diagram on the next page shows the percentage of the different kinds of vertebrates.

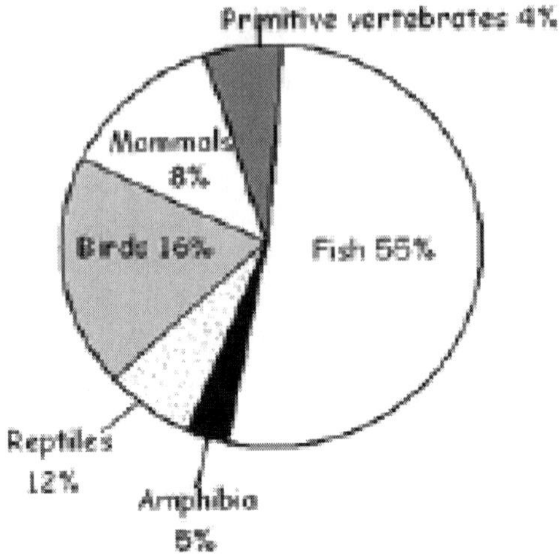

Different Kinds of Vertebrate

However, the term animal includes much more than just the Vertebrates. In fact this group makes up only a very small portion of all animals. Take a look at the diagram below, which shows the size of the different groups of animals in the Animal Kingdom as proportions of the total number of different animal species. Notice the small size of the segment representing vertebrates! All the other animals in the Animal Kingdom are animals with no backbone, or invertebrates. This includes the worms, sea anemones, starfish, snails, crabs, spiders and insects. As more than 90% of the invertebrates are insects, no wonder people worry that insects may take over the world one day!

The Classification of Vertebrates

As we have seen above the Vertebrates are divided into 5 groups or classes namely:

- Fish
- Amphibia (frogs and toads)
- Reptiles (snakes and crocodiles)
- Birds
- Mammals

These classes are all based on similarities. For instance all mammals have a similar skeleton, hair on their bodies, are warm bodied and suckle their young.

The class Mammalia (the mammals) contains 3 subclasses:

- Duck billed platypus and the spiny anteater
- Marsupials (animals like the kangaroo with pouches)
- True mammals (with a placenta)

Within the subclass containing the true mammals, there are groupings called orders that contain mammals that are more closely similar or related, than others. Examples of six mammalian orders are given below:

- Rodents (Rodentia) (rats and mice)
- Carnivores (Carnivora) (cats, dogs, bears and seals)

- Even-toed grazers (Artiodactyla) (pigs, sheep, cattle, antelopes)
- Odd-toed grazers (Perissodactyla) (horses, donkeys, zebras)
- Marine mammals (Cetacea) (whales, sea cows)
- Primates (monkeys, apes, humans)

Within each order there are various families. For example within the carnivore mammals are the families:

- Canidae (dog-like carnivores)
- Felidae (cat-like carnivores)

Even at this point it is possible to find groupings that are more closely related than others. These groups are called genera (singular genus). For instance within the cat family Felidae is the genus Felis containing the cats, as well as genera containing panthers, lynxes, and sabre toothed tigers!

The final groups within the system are the species. The definition of a species is a group of animals that can mate successfully and produce fertile offspring. This means that all domestic cats belong to the species Felis domesticus, because all breeds of cat whether Siamese, Manx or ordinary House hold cat can cross breed. However, domestic cats can not mate successfully with lions, tigers or jaguars, so these are placed in separate species, e.g. Felis leo, Felis tigris and Felis onca.

Even within the same species, there can be animals with quite wide variations in appearance that still breed successfully. We call these different breeds, races or varieties. For example there are many different breeds of dogs from Dalmatian to Chihuahua and of cats, from Siamese to Manx and domestic short-hairs, but all can cross breed. Often these breeds have been produced by selective breeding but varieties can arise in the wild when groups of animals are separated by a mountain range or sea and have developed different characteristics over long periods of time.

To summarise, the classification system consists of:

The Animal Kingdom which is divided into
Phyla which are divided into
Classes which are divided into
Orders which are divided into
Families which are divided into
Genera which are divided into
Species.

The Cell

The cell is the basic building block of living organisms. Bacteria and the parasite that causes malaria consist of single cells, while plants and animals are made up of trillions of cells. Most cells are spherical or cube shaped but some are a range of different shapes (3.1).

Most cells are so small that a microscope is needed to see them, although a few cells, e.g. the ostrich's egg, are so large that they could make a meal for several people.

A normal cell is about 0.02 of a millimetre (0.02mm) in diameter. (Small distances like this are normally expressed in micrometres or microns (?m). Note there are 1000 ?ms in every mm).

When you look at a typical animal cell with a light microscope it seems quite simple with only a few structures visible.

Three main parts can be seen:

- an outer cell wall or plasma membrane,
- an inner region called the cytoplasm and
- the nucleus

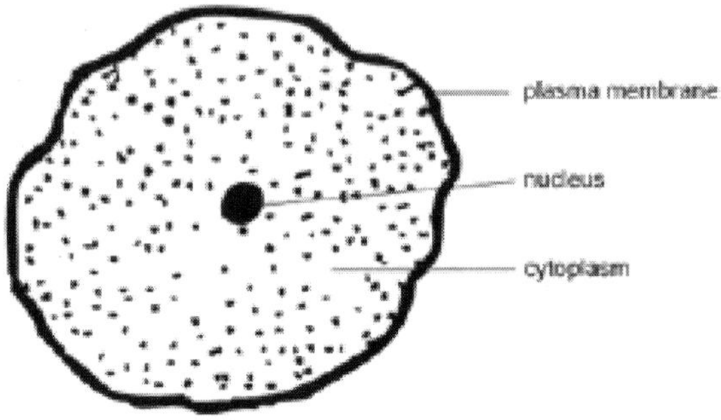

Fig: An animal cell

However, when you use an electron microscope to increase the magnification many thousands of times you see that these seemingly simple structures are incredibly complex, each with its own specialized function. For example the plasma membrane is seen to be a double layer and the cytoplasm contains many special structures called organelles (meaning little organs) which are described below.

The Plasma Membrane

The thin plasma membrane surrounds the cell, separating its contents from the surroundings and controlling what enters and leaves the cell. The plasma membrane is composed of two main molecules, fats (in fact phospholipids) and proteins. The fats are arranged in a double layer with the large protein molecules dotted about in the membrane . Some of the protein molecules form tiny channels in the membrane while others help transport substances from one side of the membrane to the other.

How substances move across the Plasma Membrane

Substances need to pass through the membrane to enter or leave the cell and they do so in a number of ways. Some of these processes require no energy i.e. they are passive, while others require energy i.e. they are active.

Passive processes include: a) diffusion and b) osmosis, while active processes include: c) active transport, d) phagocytosis, e) pinocytosis and f) exocytosis. These will be described below.

(a) Diffusion

Although you may not know it, you are already familiar with the process of diffusion. It is diffusion that causes a smell (expensive perfume or smelly socks) in one part of the room to gradually move through the room so it can be smelt on the other side. Diffusion occurs in the air and in liquids. Shows what happens when a few crystals of a dark purple dye called potassium permanganate are dropped into a beaker of water. The dye molecules diffuse into the water moving from high to low concentrations so they become evenly distributed throughout the beaker.

In the body, diffusion causes molecules that are in a high concentration on one side of the cell membrane to move across the membrane until they are present in equal concentrations on both sides. It takes place because all molecules have an in-built vibration that causes them to move and collide until they are evenly distributed. It is an absolutely natural process that requires no added energy.

Small molecules like oxygen, carbon dioxide, water and ammonia as well as fats, diffuse directly through the double fat layer of the membrane. The small molecules named above as well as a variety of charged particles (ions) also diffuse through the protein-lined channels. Larger molecules like glucose attach to a carrier molecule that aids their diffusion through the membrane. This is called facilitated diffusion.

In the animal's body diffusion is important for moving oxygen and carbon dioxide between the lungs and the blood, for moving digested food molecules from the gut into the blood and for the removal of waste products from the cell.

b) Osmosis

Although the word may be unfamiliar, you are almost certainly acquainted with the effects of osmosis. It is osmosis that plumps out dried fruit when you soak it before making a fruit cake or makes that wizened old carrot look almost like new when you soak it in water. Osmosis is in fact the diffusion of water across a membrane that allows water across but not larger molecules. This kind of membrane is called a semi-permeable membrane.

It shows a container divided into two parts by an artificial semi-permeable membrane. Water is poured into one part while a solution containing salt is poured into the other part. Water can cross the membrane but the salt cannot. The water crosses the semi-permeable membrane by diffusion until there is an equal amount of water on both sides of the membrane. The effect of this would be to make the salt solution more diluted and cause the level of the liquid in the right-hand side of the container to rise so it looked like side B of diagram 3.6. This movement of water across the semi-permeable membrane is called osmosis. It is a completely natural process that requires no outside energy.

Although it would be difficult to do in practice, imagine that you could now take a plunger and push down on the fluid in the right-hand side of container B so that it flowed back across the semi-permeable membrane until the level of fluid on both sides was equal again. If you could measure the pressure required to do this, this would be equal to the osmotic pressure of the salt solution. (This is a rather advanced concept at this stage but you will meet this term again when you study fluid balance later in the course).

The plasma membrane of cells acts as a semi-permeable membrane. If red blood cells, for example, are placed in water, the water crosses the membrane to make the amount of water on both sides of it equal . This means that the water moves into the cell causing it to swell. This can occur to such an extent that the cell actually bursts to release its contents. This bursting of red blood cells is called haemolysis. In a situation such as this when the solution on one side of a semi-permeable membrane has a lower concentration than that on the other side, the first solution is said to be hypotonic to the second.

Now think what would happen if red blood cells were placed in a salt solution that has a higher salt concentration than the solution within the cells . Such a bathing solution is called a hypertonic solution. In this situation the "concentration" of water within the cells would be higher than that outside the cells. Osmosis (diffusion of water) would then occur from the inside of the cells to the outside solution, causing the cells to shrink.

A solution that contains 0.9% salt has the same concentration as body fluids and the solution within red cells. Cells placed in such a solution would neither swell nor shrink . This solution is called an isotonic solution. This strength of salt solution is often called normal saline and is used when replacing an animal's body fluids or when cells like red blood cells have to be suspended in fluid.

Remember - osmosis is a special kind of diffusion. It is the diffusion of water molecules across a semi-permeable membrane. It is a completely passive process and requires no energy.

Sometimes it is difficult to remember which way the water molecules move. Although it is not strictly true in a biological sense, many students use the phrase "SALT SUCKS" to help them remember which way water moves across the membrane when there are two solutions of different salt concentrations on either side.

As we have seen water moves in and out of the cell by osmosis. All water movement from the intestine into the blood system and between the blood capillaries and the fluid around the cells (tissue or extra cellular fluid) takes place by osmosis. Osmosis is also important in the production of concentrated urine by the kidney.

c) Active transport

When a substance is transported from a low concentration to a high concentration i.e. uphill against the concentration gradient, energy has to be used. This is called active transport.

Active transport is important in maintaining different concentrations of the ions sodium and potassium on either side of the nerve cell membrane. It is also important for removing valuable molecules such as glucose, amino acids and sodium ions from the urine.

d) Phagocytosis

Phagocytosis is sometimes called "cell eating". It is a process that requires energy and is used by cells to move solid particles like bacteria across the plasma membrane. Once within the cell, enzymes produced by the lysosomes of the cell (described later) destroy the bacteria.

The destruction of bacteria and other foreign substance by white blood cells by the process of phagocytosis is a vital part of the defense mechanisms of the body.

e) Pinocytosis

Pinocytosis or "cell drinking" is a very similar process to phagocytosis but is used by cells to move fluids across the plasma membrane. Most cells carry out pinocytosis.

f) Exocytosis

Exocytosis is the process by means of which substances formed in the cell are moved through the plasma membrane into the fluid outside the cell (or extra-cellular fluid). It occurs in all cells but is most important in secretory cells (e.g. cells that produce digestive enzymes) and nerve cells.

The Cytoplasm

Within the plasma membrane is the cytoplasm. It consists of a clear jelly-like fluid called the a) cytosol or intracellular fluid in which b) cell inclusions, c) organelles and d) microfilaments and microtubules are found.

a) Cytosol

The cytosol consists mainly of water in which various molecules are dissolved or suspended. These molecules include proteins, fats and carbohydrates as well as sodium, potassium, calcium and chloride ions. Many of the reactions that take place in the cell occur in the cytosol.

b) Cell inclusions

These are large particles of fat, glycogen and melanin that have been produced by the cell. They are often large enough to be seen with the light microscope. For example the cells of adipose tissue (as in the insulating fat layer under the skin) contain fat that takes up most of the cell.

c) Organelles

Organelles are the "little organs" of the cell - like the heart, kidney and liver are the organs of the body. They are structures with characteristic appearances and specific "jobs" in the cell. Most can not be seen with the light microscope and so it was only when the electron microscope was developed that they were discovered. The main organelles in the cell are the ribosomes, endoplasmic reticulum, mitochondrion, Golgi complex and lysosomes.

RIBOSOMES

Ribosomes are tiny spherical organelles that make proteins by joining amino acids together. Many ribosomes are found free in the cytosol, while others are attached to the rough endoplasmic reticulum.

Endoplasmic reticulum

The endoplasmic reticulum (ER) is a network of membranes that form channels throughout the cytoplasm from the nucleus to the plasma membrane. Various molecules are made in the ER and transported around the cell in its channels. There are two types of ER: smooth ER and rough ER.

Smooth ER is where the fats in the cell are made and in some cells, where chemicals like alcohol, pesticides and carcinogenic molecules are inactivated.

The Rough ER has ribosomes attached to its surface. The function of the Rough ER is therefore to make proteins that are modified stored and transported by the ER (3.11).

Mitochondria

Mitochondria (singular mitochondrion) are oval or rod shaped organelles scattered throughout the cytoplasm. They consist of two membranes, the inner one of which is folded to increase its surface area.

Mitochondria are the "power stations" of the cell. They make energy by "burning" food molecules like glucose. This process is called cellular respiration. The reaction requires oxygen and produces carbon dioxide which is a waste product. The process is very complex and takes place in a large number of steps but the overall word equation for cellular respiration is-

Glucose + oxygen = carbon dioxide + water + energy

Note that cellular respiration is different from respiration or breathing. Breathing is the means by which air is drawn into and expelled from the lungs. Breathing is necessary to supply the cells with the oxygen required by the mitochondria and to remove the carbon dioxide produced as a waste product of cellular respiration.

Active cells like muscle, liver, kidney and sperm cells have large numbers of mitochondria.

Golgi Apparatus

The Golgi bodies in a cell together make up the Golgi apparatus. Golgi bodies are found near the nucleus and consist of flattened membranes stacked on top of each other rather like a pile of plates. The Golgi apparatus modifies and sorts the proteins and fats made by the ER, then surrounds them in a membrane as vesicles so they can be moved to other parts of the cell.

Lysosomes

Lysosomes are large vesicles that contain digestive enzymes. These break down bacteria and other substances that are brought into the cell by phagocytosis or pinocytosis. They also digest worn-out or damaged organelles, the components of which can then be recycled by the cell to make new structures.

d) Microfilaments And Microtubules

Some cells can move and change shape and organelles and chemicals are moved around the cell. Threadlike structures called microfilaments and microtubules that can contract are responsible for this movement.

These structures also form the projections from the plasma membrane known as flagella (singular flagellum) as in the sperm tail, and cilia found lining the respiratory tract and used to remove mucus that has trapped dust particles (4).

Microtubules also form the pair of cylindrical structures called centrioles found near the nucleus. These help organise the spindle used in cell division.

The Nucleus

The nucleus is the largest structure in a cell and can be seen with the light microscope. It is a spherical or oval body that contains the chromosomes. The nucleus controls the development and activity of the cell. Most cells contain a nucleus although mature red blood cells have lost theirs during development and some muscle cells have several nuclei.

A double membrane similar in structure to the plasma membrane surrounds the nucleus (now called the nuclear envelope). Pores in this nuclear membrane allow communication between the nucleus and the cytoplasm.

Within the nucleus one or more spherical bodies of darker material can be seen, even with the light microscope. These are called nucleoli and are made of RNA. Their role is to make new ribosomes.

Chromosomes

Inside the nucleus are the chromosomes which:

- contain DNA;
- control the activity of the cell;
- are transmitted from cell to cell when cells divide;
- are passed to a new individual when sex cells fuse together in sexual reproduction.

In cells that are not dividing the chromosomes are very long and thin and appear as dark grainy material. They become visible just before a cell divides when they shorten and thicken and can then be counted.

The number of chromosomes in the cells of different species varies but is constant in the cells of any one species (e.g. horses have 64 chromosomes, cats have 38 and humans 46). Chromosomes occur in pairs (i.e. 32 pairs in the horse nucleus and 19 in that of the cat). Members of each pair are identical in length and shape and if you look carefully you may be able to see some of the pairs in the human set of chromosomes.

CELL DIVISION

Cells divide when an animal grows, when its body repairs an injury and when it produces sperm and eggs (or ova). There are two types of cell division: Mitosis and meiosis.

Mitosis. This is the cell division that occurs when an animal grows and when tissues are repaired or replaced. It produces two new cells (daughter cells) each with a full set of chromosomes that are identical to each other and to the parent cell. All the cells of an animal's body therefore contain identical DNA.

Meiosis. This is the cell division that produces the ova and sperm necessary for sexual reproduction. It only occurs in the ovary and testis.

The most important function of meiosis it to halve the number of chromosomes so that when the sperm fertilises the ovum the normal number is regained. Body cells with the full set of chromosomes are called diploid cells, while gametes (sperm and ova) with half the chromosomes are called haploid cells.

Meiosis is a more complex process than mitosis as it involves two divisions one after the other and the four cells produced are all genetically different from each other and from the parent cell.

This fact that the cells formed by meiosis are all genetically different from each other and from the parent cell can be seen in litters of kittens where all the members of the litter are different from each other as well as being different from the parents although they display characteristics of both.

The Cell As A Factory

To make the function of the parts of the cell easier to understand and remember you can compare them to a factory. For example:

- The nucleus (1) is the managing director of the factory consulting the blueprint (the chromosomes) (2);
- The mitochondria (3) supply the power
- The ribosomes (4) make the products;
- The chloroplasts of plant cells (5) supply the fuel (food)

- The Golgi apparatus (6) packages the products ready for dispatch;
- The ER (7) modifies, stores and transports the products around the factory;
- The plasma membrane is the factory wall and the gates (8);
- The lysosomes dispose of the waste and worn-out machinery.

BODY ORGANISATION

Living organisms move, feed, respire (burn food to make energy), grow, sense their environment, excrete and reproduce. These seven characteristics are sometimes summarized by the words "MRS GREN". functions of:

- Movement
- Respiration
- Sensitivity
- Growth
- Reproduction
- Excretion
- Nutrition

Living organisms are made from cells which are organised into tissues and these are themselves combined to form organs and systems.

Skin cells, muscle cells, skeleton cells and nerve cells, for example. These different types of cells are not just scattered around randomly but similar cells that perform the same function are arranged in groups. These collections of similar cells are known as tissues.

There are four main types of tissues in animals. These are:

- Epithelial tissues that form linings, coverings and glands,
- Connective tissues for transport and support
- Muscle tissues for movement and
- Nervous tissues for carrying messages.

Epithelial Tissues

Epithelium (plural epithelia) is tissue that covers and lines. It covers an organ or lines a tube or space in the body. There are several different types of epithelium, distinguished by the different shapes of the cells and whether they consist of only a single layer of cells or several layers of cells.

Simple Epithelia - with a single layer of cells

Squamous epithelium

Squamous epithelium consists of a single layer of flattened cells that are shaped rather like 'crazy paving'. It is found lining the heart, blood vessels, lung alveoli and body cavities. Its thinness allows molecules to diffuse across readily.

Cuboidal epithelium

Cuboidal epithelium consists of a single layer of cube shaped cells. It is rare in the body but is found lining kidney tubules. Molecules pass across it by diffusion, osmosis and active transport.

Columnar epithelium

Columnar epithelium consists of column shaped cells. It is found lining the gut from the stomach to the anus. Digested food products move across it into the blood stream.

Columnar epithelium with cilia

Columnar epithelium with cilia on the free surface (also known as the apical side of the cell) lines the respiratory tract, fallopian tubes and uterus. The cilia beat rhythmically to transport particles.

Transitional epithelium - with a variable number of layers

The cells in transitional epithelium can move over one another allowing it to stretch. It is found in the wall of the bladder.

Stratified epithelia - with several layers of cells

Epithelia with several layers of cells are found where toughness and resistance to abrasion are needed.

Stratified squamous epithelium

Stratified squamous epithelium has many layers of flattened cells. It is found lining the mouth, cervix and vagina. Cells at the base divide and push up the cells above them and cells at the top are worn or pushed off the surface (4.6). This type of epithelium protects underlying layers and repairs itself rapidly if damaged.

Keratinised stratified squamous epithelium

Keratinised stratified squamous epithelium has a tough waterproof protein called keratin deposited in the cells. It forms the skin found covering the outer surface of mammals. .

Connective Tissues

Blood, bone, tendons, cartilage, fibrous connective tissue and fat (adipose) tissue are all classed as connective tissues. They are tissues that are used for supporting the body or transporting substances around the body. They also consist of three parts: they all have cells suspended in a ground substance or matrix and most have fibres running through it.

Blood

Blood consists of a matrix - plasma, with several types of cells and cell fragments suspended in it. The fibres are only evident in blood that has clotted. Blood will be described in detail in chapter 8.

Lymph

Lymph is similar in composition to blood plasma with various types of white blood cell floating in it. It flows in lymphatic vessels.

Connective tissue 'proper'

Connective tissue 'proper' consists of a jelly-like matrix with a dense network of collagen and elastic fibres and various cells embedded in it. There are various different forms of 'proper' connective tissue.

Loose connective tissue

Loose connective tissue is a sticky whitish substance that fills the spaces between organs. It is found in the dermis of the skin.

Dense connective tissue

Dense connective tissue contains lots of thick fibres and is very strong. It forms tendons, ligaments and heart valves and covers bones and organs like the kidney and liver.

Adipose tissue

Adipose tissue consists of cells filled with fat. It forms the fatty layer under the dermis of the skin, around the kidneys and heart and the yellow marrow of the bones.

Cartilage

Cartilage is the 'gristle' of the meat. It consists of a tough jelly-like matrix with cells suspended in it. It may contain collagen and elastic fibres. It is a flexible but tough tissue and is found at the ends of bones, in the nose, ear and trachea and between the vertebrae.

Bone

Bone consists of a solid matrix made of calcium salts that give it its hardness. Collagen fibres running through it give it its strength. Bone cells are found in spaces in the matrix. Two types of bone are found in the skeleton namely spongy and compact bone. They differ in the way the cells and matrix are arranged.

Muscle Tissues

Muscle tissue is composed of cells that contract and move the body. There are three types of muscle tissue:

Smooth muscle

Smooth muscle consists of long and slender cells with a central nucleus. It is found in the walls of blood vessels, airways to the lungs and the gut. It changes the size of the blood vessels and helps move food and fluid along. Contraction of smooth muscle fibres occurs without the conscious control of the animal.

Skeletal muscle

Skeletal muscle (sometimes called striated, striped or voluntary muscle) has striped fibres with alternating light and dark bands. It is attached to bones and is under the voluntary control of the animal.

Cardiac muscle

Cardiac muscle is found only in the walls of the heart where it produces the 'heart beat'. Cardiac muscle cells are branched cylinders with central nuclei and faint stripes. Each fibre contracts automatically but the heart beat as a whole is controlled by the pacemaker and the involuntary autonomic nervous system.

Nervous Tissues

Nervous tissue forms the nerves, spinal cord and brain. Nerve cells or neurons consist of a cell body and a long thread or axon that carries the nerve impulse. An insulating sheath of fatty material (myelin) usually surrounds the axon. 4.12 shows a typical motor neuron that sends messages to muscles to contract.

Vertebrate Bodies

We are so familiar with animals with backbones (i.e. vertebrates) that it seems rather unnecessary to point out that the body is divided into three sections. There is a well-defined head that contains the brain, the major sense organs and the mouth, a trunk that contains the other organs and a well-developed tail. Other features of vertebrates may be less apparent. For instance, vertebrates that live on the land have developed a flexible neck that is absent in fish where it would be in the way of the gills and interfere with streamlining. Mammals but not other vertebrates have a sheet of muscle called the diaphragm that divides the trunk into the chest region or thorax and the abdomen.

Body Cavities

In contrast to many primitive animals, vertebrates have spaces or body cavities that contain the body organs. Most vertebrates have a single body cavity but in mammals the diaphragm divides the main cavity into a thoracic and an abdominal cavity. In the thoracic cavity the heart and lungs are surrounded by their own membranes so that cavities are created around the heart - the pericardial cavity, and around the lungs - the pleural cavity.

Organs

Just as the various parts of the cell work together to perform the cell's functions and a large number of similar cells make up a tissue, so many different tissues can "cooperate" to form an organ that performs a particular function. For example, connective tissues, epithelial tissues, muscle tissue and nervous tissue combine to make the organ that we call the stomach. In turn the stomach combines with other organs like the intestines, liver and pancreas to form the digestive system.

Generalised Plan Of The Mammalian Body

At this point it would be a good idea to make yourself familiar with the major organs and their positions in the body of a mammal like the rabbit.

Body Systems

Organs do not work in isolation but function in cooperation with other organs and body structures to bring about the MRS GREN functions necessary to keep an animal alive. For example the stomach can only work in conjunction with the mouth and oesophagus (gullet). These provide it with the food it breaks down and digests. It then needs to pass the food on to the intestines etc. for further digestion and absorption. The organs involved with the taking of food into the body, the digestion and absorption of the food and elimination of waste products are collectively known as the digestive system.

The 11 body systems

1. Skin

The skin covering the body consists of two layers, the epidermis and dermis. Associated with these layers are hairs, feathers, claws, hoofs, glands and sense organs of the skin.

2. Skeletal System

This can be divided into the bones of the skeleton and the joints where the bones move over each other.

3. Muscular System

The muscles, in conjunction with the skeleton and joints, give the body the ability to move.

4. Cardiovascular System

This is also known as the circulatory system. It consists of the heart, the blood vessels and the blood. It transports substances around the body.

5. Lymphatic System

This system is responsible for collecting and "cleaning" the fluid that leaks out of the blood vessels. This fluid is then returned to the blood system. The lymphatic system also makes antibodies that protect the body from invasion by bacteria etc. It consists of lymphatic vessels, lymph nodes, the spleen and thymus glands.

6. Respiratory System

This is the system involved with bringing oxygen in the air into the body and getting rid of carbon dioxide, which is a waste product of processes that occur in the cell. It is made up of the trachea, bronchi, bronchioles, lungs, diaphragm, ribs and muscles that move the ribs in breathing.

7. Digestive System

This is also known as the gastrointestinal system, alimentary system or gut. It consists of the digestive tube and glands like the liver and pancreas that produce digestive secretions. It is concerned with breaking down the large molecules in foods into smaller ones that can be absorbed into the blood and lymph. Waste material is also eliminated by the digestive system.

8. Urinary System

This is also known as the renal system. It removes waste products from the blood and is made up of the kidneys, ureters and bladder.

9. Reproductive System

This is the system that keeps the species going by making new individuals. It is made up of the ovaries, uterus, vagina and fallopian tubes in the female and the testes with associated glands and ducts in the male.

10. Nervous System

This system coordinates the activities of the body and responses to the environment. It consists of the sense organs (eye, ear, semicircular canals, and organs of taste and smell), the nerves, brain and spinal cord.

11. Endocrine System

This is the system that produces chemical messengers or hormones. It consists of various endocrine glands (ductless glands) that include the pituitary, adrenal, thyroid and pineal glands as well as the testes and ovary.

HOMEOSTASIS

All the body systems, except the reproductive system, are involved with keeping the conditions inside the animal more or less stable. This is called homeostasis. These constant conditions are essential for the survival and proper functioning of the cells, tissues and organs of the body. The skin, for example, has an important role in keeping the temperature of the body constant. The kidneys keep the concentration of salts in the blood within limits and the islets of Langerhans in the pancreas maintain the correct level of glucose in the blood through the hormone insulin. As long as the various body processes remain within normal limits, the body functions properly and is healthy. Once homeostasis is disturbed disease or death may result.

Directional Terms

In the following chapters the systems of the body in the list above will be covered one by one. For each one the structure of the organs involved will be described and the way they function will be explained.

In order to describe structures in the body of an animal it is necessary to have a system for describing the position of parts of the body in relation to other parts. For example it may be necessary to describe the position of the liver in relation to the diaphragm, or the heart in relation to the lungs. Certainly if you work further with animals, in a veterinary clinic for example, it will be necessary to be able to accurately describe the position of an injury. The terms used for this are called directional terms.

The most common directional terms are right and left. However, even these are not completely straightforward especially when looking at diagrams of animals. The convention is to show the left side of the animal or organ on the right side of the page. This is the view you would get looking down on an animal lying on its back during surgery or in a post-mortem. Sometimes it is useful to imagine 'getting inside' the animal (so to speak) to check which side is which. The other common and useful directional terms are listed below and shown in diagram 4.16.

Note that we don't use the terms superior and inferior for animals. They are only used to describe the position of structures in the human body (and possibly apes) where the upright posture means some structures are above or superior to others.

In order to look at the structure of some of the parts or organs of the body it may be necessary to cut them open or even make thin slices of them that they can be examined under the microscope. The direction and position of slices or sections through an animal's body have their own terminology.

If an animal or organ is sliced lengthwise this section is called a longitudinal or sagittal section. This is sometimes abbreviated to LS.

If the section is sliced crosswise it is called a transverse or cross section. This is sometimes abbreviated to TS or XS.

The Skin

Skin comes in all kinds of textures and forms. There is the dry warty skin of toads and crocodiles, the wet slimy skin of fish and frogs, the hard shell of tortoises and the soft supple skin of snakes and humans. Mammalian skin is covered with hair, that of birds with feathers, and fish and reptiles have scales. Pigment in the skin, hairs or feathers can make the outer surface almost any colour of the rainbow.

As humans, it is often the skin of an animal that gives it its appeal to us or repels us. We love the soft feel of a cat's coat but perhaps can't bear to touch a snake. As the main part of an animal visible to us, the skin can often give us clues to the health of an animal. A healthy animal will have a clean, glowing, flexible skin, while ill health may show itself as an abnormal colour or texture.

Skin is one of the largest organs of the body, making up 6-8% of the total body weight. It consists of two distinct layers. The top layer is called the epidermis and under that is the dermis.

The epidermis is the layer that bubbles up when we have a blister and as we know from this experience, it has no blood or nerves in it. The cells at the base of the epidermis continually divide and push the cells above them upwards. As these cells move up they die and become the dry flaky scales that fall off the skin surface. The cells in the epidermis die because a special protein called keratin is deposited in them. Keratin is an extremely important substance for it makes the skin waterproof. Without it land vertebrates like reptiles, birds and mammals would, like frogs, be able to survive only in damp places.

Skin Structures Made of Keratin

Claws, Nails and Hoofs

Reptiles, birds and mammals all have nails or claws on the ends of their toes. They protect the end of the toe and may be used for grasping, grooming, digging or in defense. They are continually worn away and grow continuously from a growth layer at their base.

Hoofs are found in sheep, cows, horses etc. otherwise known as ungulate mammals. These are animals that have lost toes in the process of evolution and walk on the "nails" of the remaining toes. The hoof is a cylinder of horny material that surrounds and protects the tip of the toe.

Horns And Antlers

True horns are made of keratin and are found in sheep, goats and cattle. They are never branched and, once grown, are never shed. They consist of a core of bone arising in the dermis of the skin and are fused with the skull. The horn itself forms as a hollow cone-shaped sheath around the bone.

The antlers of male deer have quite a different structure. They are not formed in the epidermis and do not consist of keratin but are entirely of bone. They are shed each year and are often branched, especially in older animals. When growing they are covered in skin called velvet that forms the bone. Later the velvet is shed to leave the bony antler. The velvet is often removed artificially to be sold in Asia as a traditional medicine.

Other animals have projections on their heads that are not true horns either. The horns on the head of giraffes are made of bone covered with skin and hair, and the 'horn' of a rhinoceros is made of modified and fused hair-like structures.

Hair

Hair is also made of keratin and develops in the epidermis. It covers the body of most mammals where it acts as an insulator and helps to regulate the temperature of the body (see below). The colour in hairs is formed from the same pigment, melanin that colours the skin. Coat colour may help camouflage animals and sometimes acts to attract the opposite sex.

Hairs lie in a follicle and grow from a root that is well supplied with blood vessels. The hair itself consists of layers of dead keratin - containing cells and usually lies at a slant in the skin. A small bundle of smooth muscle fibres (the hair erector muscle) is attached to the side of each hair and when this contracts the hair stands on end. This increases the insulating power of the coat and is also used by some animals to make them seem larger when confronted by a foe or a competitor.

The whiskers of cats and the spines of hedgehogs are examples of special types of hairs.

Feathers

The lightness and stiffness of keratin is also a key to bird flight. In the form of feathers it provides the large airfoils necessary for flapping and gliding flight. In another form, the light fluffy down feathers,also made of keratin, are some of the best natural insulators known. This superior insulation is necessary to help maintain the high body temperatures of birds.

ountour feathers are large feathers that cover the body, wings and tail. They have an expanded vane that provides the smooth, continuous surface that is required for effective flight. This surface is formed by barbs that extend out from the central shaft. If you look carefully at a feather you can see that on either side of each barb are thousands of barbules that lock together by a complex system of hooks and notches. if this arrangement becomes disrupted, the bird uses its beak to draw the barbs and barbules together again in an action known as preening.

Down feathers are the only feathers covering a chick and form the main insulation layer under the contour feathers of the adult. They have no shaft but consist of a spray of simple, slender branches (5.8).

Pin feathers have a slender hair-like shaft often with a tiny tuft of barbs on the end. They are found between the other feathers and help tell a bird how its feathers are lying.

Skin Glands

Glands are organs that produce and secrete fluids. They are usually divided into two groups depending upon whether or not they have channels or ducts to carry their products away. Glands with ducts are called exocrine glands and include the glands found in the skin as well as the glands that produce digestive enzymes in the gut. Endocrine glands have no ducts and release their products (hormones) directly into the blood stream. The pituitary and adrenal glands are examples of endocrine glands.

Most vertebrates have exocrine glands in the skin that produce a variety of secretions. The slime on the skin of fish and frogs is mucus produced by skin glands and some fish and frogs also produce poison from modified glands. In fact the skin glands of some frogs produce the most poisonous chemicals known. Reptiles and birds have a dry skin with few glands. The preen gland, situated near the base of the bird's tail, produces oil to help keep the feathers in good condition. Mammals have an array of different skin glands. These include the wax producing, sweat, sebaceous and mammary glands.

Wax producing glands are found in the ears. Sebaceous glands secrete an oily secretion into the hair follicle. This secretion, known as sebum, keeps the hair supple and helps prevent the growth of bacteria.

Sweat glands consist of a coiled tube and a duct leading onto the skin surface. Their appearance when examined under the microscope inspired one of the first scientists to observe them to call them "fairies' intestines". Sweat contains salt and waste products like urea and the evaporation of sweat on the skin surface is one of the major mechanisms for cooling the body of many mammals. Horses can sweat up to 30 litres of fluid a day during active exercise, but cats and dogs have few sweat glands and must cool themselves by panting. Scent in the sweat of many animals is used to mark territory or attract the opposite sex.

Mammary glands are only present in mammals. They are thought to be modified sebaceous glands and are present in both sexes but are rarely active in males (5.10). The number of glands varies from species to species. They open to the surface in well-developed nipples. Milk contains proteins, sugars, fats and salts, although the exact composition varies from one species to another.

The Skin And Sun

A moderate amount of UV in sunlight is necessary for the skin to form vitamin D. This vitamin prevents bone disorders like rickets to which animals reared indoors are susceptible. Excessive exposure to the UV in sunlight can be damaging and the pigment melanin, deposited in cells at the base of the epidermis, helps to protect the underlying layers of the skin from this damage. Melanin also colours the skin and variations in the amount of melanin produces colours from pale yellow to black.

Sunburn And Skin Cancer

Excess exposure to the sun can cause sunburn. This is common in humans, but light skinned animals like cats and pigs can also be sunburned, especially on the ears. Skin cancer can also result from excessive exposure to the sun. As holes in the ozone layer increase exposure to the sun's UV rays, so too does the rate of skin cancer in humans and animals.

The Dermis

The underlying layer of the skin, known as the dermis, is much thicker but much more uniform in structure than the epidermis (5.1). It is composed of loose connective tissue with a felted mass of collagen and elastic fibres. It is this part of the skin of cattle and pigs etc. that becomes commercial leather when treated,. The dermis is well supplied with blood vessels, so cuts and burns that penetrate down into the dermis will bleed or cause serious fluid loss. There are also numerous nerve endings and touch receptors in the dermis because, of course, the skin is sensitive to touch, pain and temperature.

When looking at a section of the skin under the microscope you can see hair follicles, sweat and sebaceous glands dipping down into the dermis. However, these structures do not originate in the dermis but are derived from the epidermis.

In the lower levels of the dermis is a layer of fat or adipose tissue (5.1). This acts as an energy store and is an excellent insulator especially in mammals like whales with little hair.

The Skin and Temperature Regulation

Vertebrates can be divided into two groups depending on whether or not they control their internal temperature. Amphibia (frogs) and reptiles are said to be"cold blooded" (poikilothermic) because their body temperature approximately follows that of the environment. Birds and mammals are said to be warm blooded (homoiothermic) because they can maintain a roughly constant body temperature despite changes in the temperature of the environment.

Heat is produced by the biochemical reactions of the body (especially in the liver) and by muscle contraction. Most of the heat lost from the body occurs via the skin. It is therefore not surprising that many of the mechanisms for controlling the temperature of the body operate here.

Reduction of Heat Loss

When an animal is in a cold environment and needs to reduce heat loss the erector muscles contract causing the hair or feathers to rise up and increase the layer of insulating air trapped by them.

Heat loss from the skin surface can also be reduced by the contraction of the abundant blood vessels that lie in the dermis. This takes blood flow to deeper levels, so reducing heat loss and causing pale skin.

Shivering caused by twitching muscles produces heat that also helps raise the body temperature.

Increase of Heat Loss

There are two main mechanisms used by animals to increase the amount of heat lost from the skin when they are in a hot environment or high levels of activity are increasing internal heat production. The first is the expansion of the blood vessels in the dermis so blood flows near the skin surface and heat loss to the environment can take place. The second is by the production of sweat from the sweat glands. The evaporation of this liquid on the skin surface produces a cooling effect.

The mechanisms for regulating body temperature are under the control of a small region of the brain called the hypothalamus. This acts like a thermostat.

Heat Loss and Body Size

The amount of heat that can be lost from the surface of the body is related to the area of skin an animal has in relation to the total volume of its body.

Small animals like mice have a very large skin area compared to their total volume. This means they tend to loose large amounts of heat and have difficulty keeping warm in cold weather. They may need to keep active just to maintain their body temperature or may hibernate to avoid the problem.

Large animals like elephants have the opposite problem. They have only a relatively small skin area in relation to their total volume and may have trouble keeping cool. This is one reason that these large animals tend to have sparse coverings of hair.

THE SKELETON

Fish, frogs, reptiles, birds and mammals are called vertebrates, a name that comes from the bony column of vertebrae (the spine) that supports the body and head. The rest of the skeleton of all these animals (except the fish) also has the same basic design with a skull that houses and protects the brain and sense organs and ribs that protect the heart and lungs and, in mammals, make breathing possible. Each of the four limbs is made to the same basic pattern. It is joined to the spine by means of a flat, broad bone called a girdle and consists of one long upper bone, two long lower bones, several smaller bones in the wrist or ankle and five digits.

The Vertebral Column

The vertebral column consists of a series of bones called vertebrae linked together to form a flexible column with the skull at one end and the tail at the other. Each vertebra consists of a ring of bone with spines (spinous process) protruding dorsally from it. The spinal cord passes through the hole in the middle and muscles attach to the spines making movement of the body possible.

The shape and size of the vertebrae of mammals vary from the neck to the tail. In the neck there are cervical vertebrae with the two top ones, the atlas and axis, being specialised to support the head and allow it to nod "Yes" and shake "No". Thoracic vertebrae in the chest region have special surfaces against which the ribs move during breathing. Grazing animals like cows and giraffes that have to support weighty heads on long necks have extra large spines on their cervical and thoracic vertebrae for muscles to attach to. Lumbar vertebrae in the loin region are usually large strong vertebrae with prominent spines for the attachment of the large muscles of the lower back. The sacral vertebrae are usually fused into one solid bone called the sacrum that sits within the pelvic girdle. Finally there are a variable number of small bones in the tail called the coccygeal vertebrae.

The Skull

The skull of mammals consists of 30 separate bones that grow together during development to form a solid case protecting the brain and sense organs. The "box "enclosing and protecting the brain is called the cranium. The bony wall of the cranium encloses the middle and inner ears, protects the organs of smell in the nasal cavity and the eyes in sockets known as orbits. The teeth are inserted into the upper and lower jaws. The lower jaw is known as the mandible. It forms a joint with the skull moved by strong muscles that allow an animal to chew. At the front of the skull is the nasal cavity, separated from the mouth by a plate of bone called the palate. Behind the nasal cavity and connecting with it are the sinuses. These are air spaces in the bones of the skull which help keep the skull as light as possible. At the base of the cranium is the foramen magnum, translated as "big hole", through which the spinal cord passes. On either side of this are two small, smooth rounded knobs or condyles that articulate (move against) the first or Atlas vertebra.

The Rib

Paired ribs are attached to each thoracic vertebra against which they move in breathing. Each rib is attached ventrally either to the sternum or to the rib in front by cartilage to form the rib cage that protects the heart and lungs. In dogs one pair of ribs is not attached ventrally at all. They are called floating ribs. Birds have a large expanded sternum called the keel to which the flight muscles (the 'breast" meat of a roast chicken) are attached.

The Forelimb

The forelimb consists of: Humerus, radius and ulna, carpals, metacarpals, digits or phalanges. The top of the humerus moves against (articulates with) the scapula at the shoulder joint. By changing the number, size and shape of the various bones, fore limbs have evolved to fit different ways of life. They have become wings for flying in birds and bats, flippers for swimming in whales, seals and porpoises, fast and efficient limbs for running in horses and arms and hands for holding and manipulating in primates.

The Hind Limb

The hind limbs have a similar basic pattern to the forelimb. They consist of: femur, tibia and fibula, tarsals, metatarsals, digits or phalanges. The top of the femur moves against (articulates with) the pelvis at the hip joint.

The Long pastern or proximal phalanx is P1, not P3. The distal phalanx or coffin bone (called hoof there) is the 3rd phalanges. The patella or kneecap is embedded in a large tendon in front of the knee. It seems to smooth the movements of the knee. The legs of the horse are highly adapted to give it great galloping speed over long distances. The bones of the leg, wrist and foot are greatly elongated and the hooves are actually the tips of the third fingers and toes, the other digits having been lost or reduced.

The Girdles

The girdles pass on the "push" produced by the limbs to the body. The shoulder girdle or scapula is a triangle of bone surrounded by the muscles of the back but not connected directly to the spine. This arrangement helps it to cushion the body when landing after a leap and gives the forelimbs the flexibility to manipulate food or strike at prey. Animals that use their forelimbs for grasping, burrowing or climbing have a well-developed clavicle or collar bone. This connects the shoulder girdle to the sternum. Animals like sheep, horses and cows that use their forelimbs only for supporting the body and locomotion have no clavicle. The pelvic girdle or hipbone attaches the sacrum and the hind legs. It transmits the force of the leg-thrust in walking or jumping directly to the spine

Categories Of Bones

People who study skeletons place the different bones of the skeleton into groups according to their shape or the way in which they develop. Thus we have long bones like the femur, radius and finger bones, short bones like the ones of the wrist and ankle, irregular bones like the vertebrae and flat bones like the shoulder blade and bones of the skull. Finally there are bones that develop in tissue separated from the main skeleton. These include sesamoid bones which include bones like the patella or kneecap that develop in tendons and visceral bones that develop in the soft tissue of the penis of the dog and the cow's heart.

BIRD SKELETONS

Although the skeleton of birds is made up of the same bones as that of mammals, many are highly adapted for flight. The most noticeable difference is that the bones of the forelimbs are elongated to act as wings. The large flight muscles make up as much as 1/5th of the body weight and are attached to an extension of the sternum called the keel. The vertebrae of the lower back are fused to provide the rigidity needed to produce flying movements. There are also many adaptations to reduce the weight of the skeleton. For instance birds have a beak rather than teeth and many of the bones are hollow.

The Structure of Long Bones

A long bone consists of a central portion or shaft and two ends called epiphyses. Long bones move against or articulate with other bones at joints and their ends have flattened surfaces and rounded protuberances (condyles) to make this possible. If you carefully examine a long bone you may also see raised or rough surfaces. This is where the muscles that move the bones are attached. You will also see holes (a hole is called a foramen) in the bone. Blood vessels and nerves pass into the bone through these. You may also be able to see a fine line at each end of the bone. This is called the growth plate or epiphyseal line and marks the place where increase in length of the bone occurred.

If you cut a long bone lengthways you will see it consists of a hollow cylinder. The outer shell is covered by a tough fibrous sheath to which the tendons are attached. Under this is a layer of hard, dense compact bone (see below). This gives the bone its strength. The central cavity contains fatty yellow marrow, an important energy store for the body, and the ends are made from honeycomb-like bony material called spongy bone. Spongy bone contains red marrow where red blood cells are made.

Compact Bone

Compact bone is not the lifeless material it may appear at first glance. It is a living dynamic tissue with blood vessels, nerves and living cells that continually rebuild and reshape the bone structure as a result of the stresses, bends and breaks it experiences. Compact bone is composed of microscopic hollow cylinders that run parallel to each other along the length of the bone. Each of these cylinders is called a Haversian system. Blood vessels and nerves run along the central canal of each Haversian system. Each system consists of concentric rings of bone material (the matrix) with minute spaces in it that hold the bone cells. The hard matrix contains crystals of calcium phosphate, calcium carbonate and magnesium salts with collagen fibres that make the bone stronger and somewhat flexible. Tiny canals connect the cells with each other and their blood supply.

Spongy Bone

Spongy bone gives bones lightness with strength. It consists of an irregular lattice that looks just like an old fashioned loofah sponge. It is found on the ends of long bones and makes up most of the bone tissue of the limb girdles, ribs, sternum, vertebrae and skull. The spaces contain red marrow, which is where red blood cells are made and stored.

Bone Growth

The skeleton starts off in the foetus as either cartilage or fibrous connective tissue. Before birth and, sometimes for years after it, the cartilage is gradually replaced by bone. The long bones increase in length at the ends at an area known as the epiphyseal plate where new cartilage is laid down and then gradually converted to bone. When an animal is mature, bone growth ceases and the epiphyseal plate converts into a fine epiphyseal line.

Broken Bones

A fracture or break dramatically demonstrates the dynamic nature of bone. Soon after the break occurs blood pours into the site and cartilage is deposited. This starts to connect the broken ends together. Later spongy bone replaces the cartilage, which is itself replaced by compact bone. Partial healing to the point where some weight can be put on the bone can take place in 6 weeks but complete healing may take 3-4 months.

Joints

Joints are the structures in the skeleton where 2 or more bones meet. There are several different types of joints. Some are immovable once the animal has reached maturity. Examples of these are those between the bones of the skull and the midline joint of the pelvic girdle. Some are slightly moveable like the joints between the vertebrae but most joints allow free movement and have a typical structure with a fluid filled cavity separating the articulating surfaces (surfaces that move against each other) of the two bones. This kind of joint is called a synovial joint. The joint is held together by bundles of white fibrous tissue called ligaments and a fibrous capsule encloses the joint. The inner layers of this capsule secrete the synovial fluid that acts as a lubricant. The articulating surfaces of the bones are covered with cartilage that also reduces friction and some joints, e.g. the knee, have a pad of cartilage between the surfaces that articulate with each other.

The shape of the articulating bones in a joint and the arrangement of ligaments determine the kind of movement made by the joint. Some joints only allow a to and from gliding movement e.g. between the ankle and wrist bones; the joints at the elbow, knee and fingers are hinge joints and allow movement in two dimensions and the axis vertebra pivots on the atlas vertebra. Ball and socket joints, like those at the shoulder and hip, allow the greatest range of movement.

Muscles

Muscles make up the bulk of an animal's body and account for about half its weight. The meat on the chop or roast is muscle and is composed mainly of protein. The cells that make up muscle tissue are elongated and able to contract to a half or even a third of their length when at rest. There are three different kinds of muscle; smooth, cardiac and skeletal muscle.

Smooth muscle

Smooth or Involuntary muscle carries out the unconscious routine tasks of the body such as moving food down the digestive system, keeping the eyes in focus and adjusting the diameter of blood vessels. The individual cells are spindle-shaped, being fatter in the middle and tapering off towards the ends with a nucleus in the centre of the cell. They are usually found in sheets and are stimulated by the non-conscious or autonomic nervous system as well as by hormones.

Cardiac muscle

Cardiac muscle is only found in the wall of the heart. It is composed of branching fibres that form a three-dimensional network. When examined under the microscope, a central nucleus and faint stripes or striations can be seen in the cells. Cardiac muscle cells contract spontaneously and rhythmically without outside stimulation but the pacemaker coordinates the heart beat. Nerves and hormones modify this rhythm.

Skeletal muscle

Skeletal muscle is the muscle that is attached to and moves the skeleton, and is under voluntary control. It is composed of elongated cells or fibres lying parallel to each other. Each cell is unusual in that it has several nuclei and when examined under the microscope appears striped or striated. This appearance gives the muscle its names of striped or striated muscle. Each cell of striated muscle contains hundreds, or even thousands, of microscopic fibres each one with its own striped appearance. The stripes are formed by two different sorts of protein that slide over each other making the cell contract.

Muscle contraction

Muscle contraction requires energy and muscle cells have numerous mitochondria. However, only about 15% of the energy released by the mitochondria is used to fuel muscle contraction. The rest is released as heat. This is why exercise increases body temperature and makes animals sweat or pant to rid themselves of this heat.

What we refer to as a muscle is made up of groups of muscle fibres surrounded by connective tissue. The connective tissue sheaths join together at the ends of the muscle to form tough white bands of fibre called tendons. These attach the muscles to the bones. Tendons are similar in structure to the ligaments that attach bones together across a joint.

Structure of a muscle

A single muscle is fat in the middle and tapers towards the ends. The middle part, which gets fatter when the muscle contracts, is called the belly of the muscle. If you contract your biceps muscle in your upper arm you may feel it getting fatter in the middle. You may also notice that the biceps is attached at its top end to bones in your shoulder while at the bottom it is attached to bones in your lower arm. Notice that the bones at only one end move when you contract the biceps. This end of the muscle is called the insertion. The other end of the muscle, the origin, is attached to the bone that moves the least

Antagonistic muscles

Skeletal muscles usually work in pairs. When one contracts the other relaxes and vice versa. Pairs of muscles that work like this are called antagonistic muscles. For example the muscles in the upper forearm are the biceps and triceps . Together they bend the elbow. When the biceps contracts (and the triceps relaxes) the lower forearm is raised and the angle of the joint is reduced. This kind of movement is called flexion. When the triceps is contracted (and the biceps relaxes), the angle of the elbow increases. The term for this movement is extension.

When you or animals contract skeletal muscle it is a voluntary action. For example, you make a conscious decision to walk across the room, raise the spoon to your mouth or smile. There is however, another way in which contraction of muscles attached to the skeleton happens that is not under voluntary control. This is during a reflex action, such as jerking your hand away from the hot stove you have touched by accident.

CARDIOVASCULAR SYSTEM

Blood

Blood is a unique fluid containing cells that is pumped by the heart around the body of animals in a system of pipes known as the circulatory system. It carries oxygen and nutrients to the cells of the body and removes waste products like carbon dioxide from them. Blood is also important for keeping conditions in the body constant, in other words for maintaining homeostasis. It helps keep the acidity or pH stable and helps maintain a constant temperature in the body. Blood also has an important role in defending the body against disease.

A simple way to find out what is in blood is to remove a small amount from an animal and place it in a tube with a substance that prevents it from clotting (an anticoagulant). If you leave the tube to stand for a few hours you will find that it settles out into two layers. The top layer consists of a light yellow fluid, the plasma, and the bottom layer consists of red blood cells (RBCs). If you look very carefully you can also see a thin beige-coloured layer in between these two layers. This consists of the white blood cells (WBCs).

The above procedure is usually done more rapidly by placing the blood sample in a centrifuge for a few minutes. This machine acts like a super spin drier rotating about 10,000 times a minute and packing the heavier particles (red blood cells) at the bottom of the tube. The sample that results is called the packed cell volume (P.C.V.) or haematocrit. It is a very useful measurement of the concentration of red blood cells in the blood. For most animals the packed cell volume is in the range 30-45%. If it is lower than this it means that the concentration of red blood cells is low and the animal is anaemic. If the reading is above this range it may mean the animal is dehydrated. Animals that live at high altitudes also have high P.C.V.s to compensate for the low oxygen concentration there.

The Heart

The heart is the pump that pushes the blood around the body in the blood vessels of the circulatory system. In fish the blood only passes through the heart once on its way to the gills and then round the rest of the body. However, in mammals and birds that have lungs, the blood passes through the heart twice: once on its way to the lungs where it picks up oxygen and then through the heart again to be pumped all over the body. The heart is therefore two separate pumps, side by side.

The heart is situated in the thorax between the lungs and is protected by the rib cage. In some animals it is displaced slightly to the left-hand side. A tough membrane called the pericardium covers it. There is a narrow space between the pericardium and the heart that is filled with a liquid that acts as a lubricant.

The heart of mammals is a hollow bag made of cardiac muscle (4). The cavity inside the heart is divided into 4 chambers. The chambers on the right side are completely separate from the chambers on the left side. The two upper chambers are thin walled, and are called the atria (or auricles). The two lower chambers are thick walled and are called the ventricles.

Blood circulation

The circulatory system is the continuous system of tubes through which the blood is pumped around the body. It supplies the tissues with their requirements and removes waste products. In mammals and birds the blood circulates through two separate systems - the first from the heart to the lungs and back to the heart again (the pulmonary circulation) and the second from the heart to the head and body and back again (the systemic circulation).

The tubes through which the blood flows are the arteries, capillaries and veins. The heart pumps blood into arteries that carry it away from the heart. The arteries divide into very thin vessels called capillaries that form a network between the cells of the body. The capillaries then join up again to make veins that return the blood to the heart.

RESPIRATORY SYSTEM

Animals require a supply of energy to survive. This energy is needed to build large molecules like proteins and glycogen, make the structures in cells, move chemicals through membranes and around cells, contract muscles, transmit nerve impulses and keep the body warm. Animals get their energy from the large molecules that they eat as food. Glucose is often the energy source but it may also come from other carbohydrates, as well as fats and protein. The energy is made by the biochemical process known as cellular respiration that takes place in the mitochondria inside every living cell.

The overall reaction can be summarised by the word equation given below.

Charbohydrate Food (glucose) + Oxygen = Carbon Dioxide + Water + energy

As you can see from this equation, the cells need to be supplied with oxygen and glucose and the waste product, carbon dioxide, which is poisonous to cells, needs to be removed. The way the digestive system provides the glucose for cellular respiration, but here we are only concerned with the two gases, oxygen and carbon dioxide, that are involved in cellular respiration. These gases are carried in the blood to and from the tissues where they are required or produced.

Oxygen enters the body from the air (or water in fish)and carbon dioxide is usually eliminated from the same part of the body. This process is called gas exchange. In fish gas exchange occurs in the gills, in land dwelling vertebrates lungs are the gas exchange organs and frogs use gills when they are tadpoles and lungs, the mouth and the skin when adults.

Mammals (and birds) are active and have relatively high body temperatures so they require large amounts of oxygen to provide sufficient energy through cellular respiration. In order to take in enough oxygen and release all the carbon dioxide produced they need a very large surface area over which gas exchange can take place. The many minute air sacs or alveoli of the lungs provide this. When you look at these under the microscope they appear rather like bunches of grapes covered with a dense network of fine capillaries. A thin layer of water covers the inner surface of each alveolus. There is only a very small distance -just 2 layers of thin cells - between the air in the alveoli and the blood in the capillaries. The gases pass across this gap by diffusion.

Diffusion And Transport Of Oxygen

The air in the alveoli is rich in oxygen while the blood in the capillaries around the alveoli is deoxygenated. This is because the haemoglobin in the red blood cells has released all the oxygen it has been carrying to the cells of the body. Oxygen diffuses from high concentration to low concentration. It therefore crosses the narrow barrier between the alveoli and the capillaries to enter the blood and combine with the haemoglobin in the red blood cells to form oxyhaemoglobin.

The narrow diameter of the capillaries around the alveoli means that the blood flow is slowed down and that the red cells are squeezed against the capillary walls. Both of these factors help the oxygen diffuse into the blood.

When the blood reaches the capillaries of the tissues the oxygen splits from the haemoglobin molecule. It then diffuses into the tissue fluid and then into the cells.

Diffusion and Transport of Carbon Dioxide

Blood entering the lung capillaries is full of carbon dioxide that it has collected from the tissues. Most of the carbon dioxide is dissolved in the plasma either in the form of sodium bicarbonate or carbonic acid. A little is transported by the red blood cells. As the blood enters the lungs the carbon dioxide gas diffuses through the capillary and alveoli walls into the water film and then into the alveoli. Finally it is removed from the lungs during breathing out.

The Air Passages

When air is breathed in it passes from the nose to the alveoli of the lungs down a series of tubes. After entering the nose the air passes through the nasal cavity, which is lined with a moist membrane that adds warmth and moisture to the air as it passes. The air then flows through the pharynx or throat, a passage that carries both food and air, to the larynx where the voice-box is located. Here the passages for food and air separate again. Food must pass into the oesophagus and the air into the windpipe or trachea. To prevent food entering this, a small flap of tissue called the epiglottis closes the opening during swallowing. A reflex that inhibits breathing during swallowing also (usually) prevents choking on food.

The trachea is the tube that ducts the air down the throat. Incomplete rings of cartilage in its walls help keep it open even when the neck is bent and head turned. The fact that acrobats and people that tie themselves in knots doing yoga still keep breathing during the most contorted manoeuvres shows how effective this arrangement is. The air passage now divides into the two bronchi that take the air to the right and left lungs before dividing into smaller and smaller bronchioles that spread throughout the lungs to carry air to the alveoli. Smooth muscles in the walls of the bronchi and bronchioles adjust the diameter of the air passages.

The tissue lining the respiratory passages produces mucus and is covered with minute hairs or cilia. Any dust that is breathed into the respiratory system immediately gets entangled in the mucous and the cilia move it towards the mouth or nose where it can be coughed up or blown out.

The Lungs And The Pleural Cavities

The lungs fill most of the chest or thoracic cavity, which is completely separated from the abdominal cavity by the diaphragm. The lungs and the spaces in which they lie (called the pleural cavities) are covered with membranes called the pleura. There is a thin film of fluid between the two membranes. This lubricates them as they move over each other during breathing movements.

Collapsed Lungs

The pleural cavities are completely airtight with no connection with the outside and if they are punctured by accident (a broken rib will often do this), air rushes in and the lung collapses. Separating the two lungs is a region of tissue that contains the oesophagus, trachea, aorta, vena cava and lymph nodes. This is called the mediastinum. In humans and sheep it separates the cavity completely so that puncturing one pleural cavity leads to the collapse of only one lung. In dogs, however, this separation is incomplete so a puncture results in a complete collapse of both lungs.

Breathing

The process of breathing moves air in and out of the lungs. Sometimes this process is called respiration but it is important not to confuse it with the chemical process, cellular respiration, that takes place in the mitochondria of cells. Breathing is brought about by the movement of the diaphragm and the ribs.

Inspiration

The diaphragm is a thin sheet of muscle that completely separates the abdominal and thoracic cavities. When at rest it domes up into the thoracic cavity but during breathing in or inspiration it flattens. At the same time special muscles in the chest wall move the ribs forwards and outwards. These movements of both the diaphragm and the ribs cause the volume of the thorax to increase. Because the pleural cavities are airtight, the lungs expand to fill this increased space and air is drawn down the trachea into the lungs.

Expiration

Expiration or breathing out consists of the opposite movements. The ribs move down and in and the diaphragm resumes its domed shape so the air is expelled. Expiration is usually passive and no energy is required (unless you are blowing up a balloon).

Lung Volumes

As you sit here reading this just pay attention to your breathing. Notice that your in and out breaths are really quite small and gentle (unless you have just rushed here from somewhere else!). Only a small amount of the total volume that your lungs hold is breathed in and out with each breath. This kind of gentle "at rest" breathing is called tidal breathing and the volume breathed in or out (they should be the same) is the tidal volume. Sometimes people want to measure the volume of air inspired or expired during a minute of this normal breathing. This is called the minute volume. It could be estimated by measuring the volume of one tidal breath and then multiplying that by the number of breaths in a minute. Of course it is possible to take a deep breath and breathe in as far as you can and then expire as far as possible. The volume of the air expired when a maximum expiration follows a maximum inspiration is called the vital capacity.

Composition of Air

The air animals breathe in consists of 21% oxygen and 0.04% carbon dioxide. Expelled air consists of 16% oxygen and 4.4% carbon dioxide. This means that the lungs remove only a quarter of the oxygen contained in the air. This is why it is possible to give someone (or an animal) artificial respiration by blowing expired air into their mouth.

Breathing is usually an unconscious activity that takes place whether you are awake or asleep, although, humans at least, can also control it consciously. Two regions in the hindbrain called the medulla oblongata and pons control the rate of breathing. These are called respiratory centres. They respond to the concentration of carbon dioxide in the blood. When this concentration rises during a bout of activity, for example, nerve impulses are automatically sent to the diaphragm and rib muscles that increase the rate and the depth of breathing. Increasing the rate of breathing also increases the amount of oxygen in the blood to meet the needs of this increased activity.

The Acidity Of The Blood And Breathing

The degree of acidity of the blood (the acid-base balance) is critical for normal functioning of cells and the body as a whole. For example, blood that is too acidic or alkaline can seriously affect nerve function causing a coma, muscle spasms, convulsions and even death. Carbon dioxide carried in the blood makes the blood acidic and the higher the concentration of carbon dioxide the more acidic it is. This is obviously dangerous so there are various mechanisms in the body that bring the acid-base balance back within the normal range. Breathing is one of these homeostatic mechanisms. By increasing the rate of breathing the animal increases the amount of dissolved carbon dioxide that is expelled from the blood. This reduces the acidity of the blood.

Breathing In Birds

Birds have a unique respiratory system that enables them to respire at the very high rates necessary for flight. The lungs are relatively solid structures that do not change shape and size in the same way as mammalian lungs do. Tubes run through them and connect with a series of air sacs situated in the thoracic and abdominal body cavities and some of the bones. Movements of the ribs and breastbone or sternum expand and compress these air sacs so they act rather like bellows and pump air through the lungs. The evolution of this extremely efficient system of breathing has enabled birds to migrate vast distances and fly at altitudes higher than the summit of Everest.

Lymphatic System

When tissue fluid enters the small blind-ended lymphatic capillaries that form a network between the cells it becomes lymph. Lymph is a clear watery fluid that is very similar to blood plasma except that it contains large numbers of white blood cells, mostly lymphocytes. It also contains protein, cellular debris, foreign particles and bacteria. Lymph that comes from the intestines also contains many fat globules following the absorption of fat from the digested food into the lymphatics (lacteals) of the villi (11 for more on these). From the lymph capillaries the lymph flows into larger tubes called lymphatic vessels. These carry the lymph back to join the blood circulation.

Lymphatic vessels

Lymphatic vessels have several similarities to veins. Both are thin walled and return fluid to the right hand side of the heart. The movement of the fluid in both is brought about by the contraction of the muscles that surround them and both have valves to prevent backflow. One important difference is that lymph passes through at least one lymph node or gland before it reaches the blood system. These filter out used cell parts, cancer cells and bacteria and help defend the body from infection.

Lymph nodes are of various sizes and shapes and found throughout the body. They consist of lymph tissue surrounded by a fibrous sheath. Lymph flows into them through a number of incoming vessels. It then trickles through small channels where white cells called macrophages (derived from monocytes) remove the bacteria and debris by engulfing and digesting them. The lymph then leaves the lymph nodes through outgoing vessels to continue its journey towards the heart where it rejoins the blood circulation.

As well as filtering the lymph, lymph nodes produce the white cells known as lymphocytes. Lymphocytes are also produced by the thymus, spleen and bone marrow. There are two kinds of lymphocyte. The first attach invading micro organisms directly while others produce antibodies that circulate in the blood and attack them.

The function of the lymphatic system can therefore be summarized as transport and defense. It is important for returning the fluid and proteins that have escaped from the blood capillaries to the blood system and is also responsible for picking up the products of fat digestion in the small intestine. Its other essential function is as part of the immune system, defending the body against infection.

Problems with lymph nodes and the lymphatic system

During infection of the body the lymph nodes often become swollen and tender because of their increased activity. This is what causes the swollen 'glands' in your neck during throat infections, mumps and tonsillitis. Sometimes the bacteria multiply in the lymph node and cause inflammation. Cancer cells may also be carried to the lymph nodes and then transported to other parts of the body where they may multiply to form a secondary growth or metastasis. The lymphatic system may therefore contribute to the spread of cancer. Inactivity of the muscles surrounding the lymphatic vessels or blockage of these vessels causes tissue fluid to 'back up' in the tissues resulting in swelling or oedema.

Other Organs of The Lymphatic System

The spleen is an important part of the lymphatic system. It is a deep red organ situated in the abdomen caudal to the stomach. It is composed of two different types of tissue. The first type makes and stores lymphocytes, the cells of the immune system. The second type of tissue destroys worn out red blood cells, breaking down the haemoglobin into iron, which is recycled, and waste products that are excreted. The spleen also stores red blood cells. When severe blood loss occurs, it contracts and releases them into the circulation. The thymus is a large pink organ lying just under the sternum (breastbone) just cranial to the heart. It has an important function processing lymphocytes so they are capable of recognising and attacking foreign invaders like bacteria.

Other lymph organs are the bone marrow of the long bones where lymphocytes are produced and lymph nodules, which are like tiny lymph nodes. Large clusters of these are found in the wall of the small intestine (called Peyer's Patches) and in the tonsils.

THE GUT AND DIGESTION

Plant cells are made of organic molecules using energy from the sun. This process is called photosynthesis. Animals rely on these ready-made organic molecules to supply them with their food. Some animals (herbivores) eat plants; some (carnivores) eat the herbivores.

Herbivores

Although herbivores eat plant material they have a basic problem. Their own digestive enzymes do not break down the large cellulose molecules in the plant cell walls. Micro-organisms like bacteria, on the other hand, can break them down. Therefore herbivores employ micro-organisms to do the job for them.

There are two types of herbivore:

The first, ruminants like cattle, sheep and goats, house these bacteria in a special compartment in the enlarged stomach called the rumen.

The second group has an enlarged large intestine and caecum, called a functional caecum, occupied by cellulose digesting micro-organisms. These non-ruminant herbivores include the horse, rabbit and rat.

Even with the help of microorganisms, eating plants still poses problems. Plants are not a good source of nutrients, and herbivores have to eat large quantities of food to obtain all they require. Herbivores like cows, horses and rabbits typically spend much of their day feeding. To give the micro- organisms access to the cellulose molecules, the plant cell walls need to be broken down. This is why herbivores have teeth that are adapted to crush and grind. Their guts also tend to be lengthy and the food takes a long time to pass through it.

Eating plants does have some advantages. Plants are immobile so herbivores normally have to spend little energy collecting them. This contrasts with the other main group of animals- the carnivores that often have to chase their prey.

Carnivores

Carnivorous animals like those in the cat and dog families, polar bears, seals, crocodiles and birds of prey catch and eat other animals. They often have to use large amounts of energy finding, stalking, catching and killing their prey. However, they are rewarded by the fact that meat provides a very concentrated source of nutrients. Carnivores in the wild therefore tend to eat distinct meals often with long and irregular intervals between them. Time after feeding is spent digesting and absorbing the food.

The guts of carnivores are usually shorter and less complex than those of herbivores because meat is easier to digest than plant material. Carnivores usually have teeth that are specialised for dealing with flesh, gristle and bone. They have sleek bodies, strong, sharp claws and keen senses of smell, hearing and sight. They are also often cunning, alert and have an aggressive nature.

OMNIVORES

Many animals feed on both animal and vegetable material - they are omnivorous. Most primates including humans belong to this category as do pigs and rats. Their food is diverse, ranging from plant material to animals they have either killed themselves or scavenged from other carnivores. Omnivores lack the specialised teeth and guts of carnivores and herbivores but are often highly intelligent and adaptable reflecting their varied diet.

Treatment of Food

Whether an animal eats plants or flesh, the carbohydrates, fats and proteins in the food it eats are generally giant molecules (1). These need to be split up into smaller ones before they can pass into the blood and enter the cells to be used for energy or to make new cell constituents.

For example:

Carbohydrates like cellulose, starch, and glycogen need to be split into glucose and other monosaccharides;

Proteins need to be split into amino acids;

Fats or lipids need to be split into fatty acids and glycerol.

The Gut

The digestive tract, alimentary canal or gut is a hollow tube stretching from the mouth to the anus. It is the organ system concerned with the treatment of foods.

At the mouth the large food molecules are taken into the gut - this is called ingestion. They must then be broken down into smaller ones by digestive enzymes - digestion, before they can be taken from the gut into the blood stream - absorption. The cells of the body can then use these small molecules - assimilation. The indigestible waste products are eliminated from the body by the act of egestion.

The food that enters the mouth passes to the oesophagus, then to the stomach, small intestine, cecum, large intestine, rectum and finally undigested material exits at the anus. The liver and pancreas produce secretions that aid digestion and the gall bladder stores bile.

Mouth

The mouth takes food into the body. The lips hold the food inside the mouth during chewing and allow the baby animal to suck on its mother's teat. In elephants the lips (and nose) have developed into the trunk which is the main food collecting tool. Some mammals, e.g. hamsters, have stretchy cheek pouches that they use to carry food or material to make their nests.

The sight or smell of food and its presence in the mouth stimulates the salivary glands to secrete saliva. There are four pairs of these glands in cats and dogs. The fluid they produce moistens and softens the food making it easier to swallow. It also contains the enzyme, salivary amylase, which starts the digestion of starch.

The tongue moves food around the mouth and rolls it into a ball for swallowing. Taste buds are located on the tongue and in dogs and cats it is covered with spiny projections used for grooming and lapping. The cow's tongue is prehensile and wraps around grass to graze it.

Swallowing is a complex reflex involving 25 different muscles. It pushes food into the oesophagus and at the same time a small flap of tissue called the epiglottis closes off the windpipe so food doesn't go 'down the wrong way' and choke the animal.

Teeth

Teeth seize, tear and grind food. They are inserted into sockets in the bone and consist of a crown above the gum and root below. The crown is covered with a layer of enamel, the hardest substance in the body. Below this is the dentine, a softer but tough and shock resistant material. At the centre of the tooth is a space filled with pulp which contains blood vessels and nerves. The tooth is cemented into the socket and in most teeth the tip of the root is quite narrow with a small opening for the blood vessels and nerves (11.5).

In teeth that grow continuously, like the incisors of rodents, the opening remains large and these teeth are called open rooted teeth. Mammals have 2 distinct sets of teeth. The first the milk teeth are replaced by the permanent teeth.

Types of Teeth

All the teeth of fish and reptiles are similar but mammals usually have four different types of teeth.

The incisors are the chisel-shaped 'biting off' teeth at the front of the mouth. In rodents and rabbits the incisors never stop growing (open-rooted teeth). They must be worn or ground down continuously by gnawing. They have hard enamel on one surface only so they wear unevenly and maintain their sharp cutting edge.

The largest incisors in the animal kingdom are found in elephants, for tusks are actually giant incisors. Sloths have no incisors at all, and sheep have no incisors in the upper jaw. Instead there is a horny pad against which the bottom incisors cut.

The canines or 'wolf-teeth' are long, cone-shaped teeth situated just behind the incisors. They are particularly well developed in the dog and cat families where they are used to hold, stab and kill the prey.

The tusks of boars and walruses are large canines while rodents and herbivores like sheep have no (or reduced) canines. In these animals the space where the canines would normally be is called the diastema. In rodents like the rat and beaver it allows the debris from gnawing to be expelled easily.

The cheek teeth or premolars and molars crush and grind the food. They are particularly well developed in herbivores where they have complex ridges that form broad grinding surfaces. These are created from alternating bands of hard enamel and softer dentine that wear at different rates.

In carnivores the premolars and molars slice against each other like scissors and are called carnassial teeth. They are used for shearing flesh and bone.

Dental Formula

The numbers of the different kinds of teeth can be expressed in a dental formula. This gives the numbers of incisors, canines, premolars and molars in one half of the mouth. The numbers of these four types of teeth in the left or right half of the upper jaw are written above a horizontal line and the four types of teeth in the right or left half of the lower jaw are written below it.

Thus the dental formula for the sheep is:

0.0.3.3

3.1.3.3

It indicates that in the upper right (or left) half of the jaw there are no incisors or canines (i.e. there is a diastema), three premolars and three molars. In the lower right (or left) half of the jaw are three incisors, one canine, three premolars and three molars.

The formula indicates that in the right (or left) half of the upper jaw there are three incisors, one canine, four premolars and two molars. In the right (or left) half of the lower jaw there are three incisors, one canine, four premolars and three molars.

OESOPHAGUS

The oesophagus transports food to the stomach. Food is moved along the oesophagus, as it is along the small and large intestines, by contraction of the smooth muscles in the walls that push the food along rather like toothpaste along a tube. This movement is called peristalsis.

Stomach

The stomach stores and mixes the food. Glands in the wall secrete gastric juice that contains enzymes to digest protein and fats as well as hydrochloric acid to make the contents very acidic. The walls of the stomach are very muscular and churn and mix the food with the gastric juice to form a watery mixture called chyme (pronounced kime). Rings of muscle called sphincters at the entrance and exit to the stomach control the movement of food into and out of it.

Small Intestine

Most of the breakdown of the large food molecules and absorption of the smaller molecules take place in the long and narrow small intestine. The total length varies but it is about 6.5 metres in humans, 21 metres in the horse, 40 metres in the ox and over 150 metres in the blue whale.

It is divided into 3 sections: the duodenum (after the stomach), jejunum and ileum. The duodenum receives 3 different secretions:

1. Bile from the liver;
2. Pancreatic juice from the pancreas and
3. Intestinal juice from glands in the intestinal wall.

These complete the digestion of starch, fats and protein. The products of digestion are absorbed into the blood and lymphatic system through the wall of the intestine, which is lined with tiny finger-like projections called villi that increase the surface area for more efficient absorption.

The Rumen

In ruminant herbivores like cows, sheep and antelopes the stomach is highly modified to act as a "fermentation vat". It is divided into four parts. The largest part is called the rumen. In the cow it occupies the entire left half of the abdominal cavity and can hold up to 270 litres. The reticulum is much smaller and has a honeycomb of raised folds on its inner surface. In the camel the reticulum is further modified to store water. The next part is called the omasum with a folded inner surface. Camels have no omasum. The final compartment is called the abomasum. This is the 'true' stomach where muscular walls churn the food and gastric juice is secreted

Ruminants swallow the grass they graze almost without chewing and it passes down the oesophagus to the rumen and reticulum. Here liquid is added and the muscular walls churn the food. These chambers provide the main fermentation vat of the ruminant stomach. Here bacteria and single-celled animals start to act on the cellulose plant cell walls. These organisms break down the cellulose to smaller molecules that are absorbed to provide the cow or sheep with energy. In the process, the gases methane and carbon dioxide are produced. These cause the "burps" you may hear cows and sheep making.

Not only do the micro-organisms break down the cellulose but they also produce the vitamins E, B and K for use by the animal. Their digested bodies provide the ruminant with the majority of its protein requirements.

In the wild grazing is a dangerous activity as it exposes the herbivore to predators. They crop the grass as quickly as possible and then when the animal is in a safer place the food in the rumen can be regurgitated to be chewed at the animal's leisure. This is 'chewing the cud' or rumination. The finely ground food may be returned to the rumen for further work by the microorganisms or, if the particles are small enough, it will pass down a special groove in the wall of the oesophagus straight into the omasum. Here the contents are kneaded and water is absorbed before they pass to the abomasum. The abomasum acts as a "proper" stomach and gastric juice is secreted to digest the protein.

Large Intestine

The large intestine consists of the caecum, colon and rectum. The chyme from the small intestine that enters the colon consists mainly of water and undigested material such as cellulose (fibre or roughage). In omnivores like the pig and humans the main function of the colon is absorption of water to give solid faeces. Bacteria in this part of the gut produce vitamins B and K.

The caecum, which forms a dead-end pouch where the small intestine joins the large intestine, is small in pigs and humans and helps water absorption. However, in rabbits, rodents and horses, the caecum is very large and called the functional caecum. It is here that cellulose is digested by micro-organisms. The appendix, a narrow dead end tube at the end of the caecum, is particularly large in primates but seems to have no digestive function.

FUNCTIONAL CAECUM

The caecum in the rabbit, rat and guinea pig is greatly enlarged to provide a "fermentation vat" for micro-organisms to break down the cellulose plant cell walls. This is called a functional caecum. In the horse both the caecum and the colon are enlarged. As in the rumen, the large cellulose molecules are broken down to smaller molecules that can be absorbed. However, the position of the functional caecum after the main areas of digestion and absorption, means it is potentially less effective than the rumen. This means that the small molecules that are produced there can not be absorbed by the gut but pass out in the faeces. The rabbit and rodents (and foals) solve this problem by eating their own faeces so that they pass through the gut a second time and the products of cellulose digestion can be absorbed in the small intestine. Rabbits produce two kinds of faeces. Softer night-time faeces are eaten directly from the anus and the harder pellets you are probably familiar with, that have passed through the gut twice.

The Gut of Birds

Birds' guts have important differences from mammals' guts. Most obviously, birds have a beak instead of teeth. Beaks are much lighter than teeth and are an adaptation for flight. Imagine a bird trying to take off and fly with a whole set of teeth in its head! At the base of the oesophagus birds have a bag-like structure called a crop. In many birds the crop stores food before it enters the stomach, while in pigeons and doves glands in the crop secretes a special fluid called crop-milk which parent birds regurgitate to feed their young. The stomach is also modified and consists of two compartments. The first is the true stomach with muscular walls and enzyme secreting glands. The second compartment is the gizzard. In seed eating birds this has very muscular walls and contains pebbles swallowed by the bird to help grind the food. This is the reason why you must always supply a caged bird with grit. In birds of prey like the falcon the walls of the gizzard are much thinner and expand to accommodate large meals.

Digestion

During digestion the large food molecules are broken down into smaller molecules by enzymes. The three most important groups of enzymes secreted into the gut are:

1. Amylases that split carbohydrates like starch and glycogen into monosaccharides like glucose.
2. Proteases that split proteins into amino acids.
3. Lipases that split lipids or fats into fatty acids and glycerol.

Glands produce various secretions which mix with the food as it passes along the gut.

These secretions include:

1. Saliva secreted into the mouth from several pairs of salivary glands. Saliva consists mainly of water but contains salts, mucous and salivary amylase. The function of saliva is to lubricate food as it is chewed and swallowed and salivary amylase begins the digestion of starch.
2. Gastric juice secreted into the stomach from glands in its walls. Gastric juice contains pepsin that breaks down protein and hydrochloric acid to produce the acidic conditions under which this enzyme works best. In baby animals rennin to digest milk is also produced in the stomach.
3. Bile produced by the liver. It is stored in the gall bladder and secreted into the duodenum via the bile duct. (Note that the horse, deer, parrot and rat have no gall bladder). Bile is not a digestive enzyme. Its function is to break up large globules of fat into smaller ones so the fat splitting enzymes can gain access the fat molecules.

ANCREATIC JUICE

The pancreas is a gland located near the beginning of the duodenum. In most animals it is large and easily seen but in rodents and rabbits it lies within the membrane linking the loops of the intestine (the mesentery) and is quite difficult to find. Pancreatic juice is produced in the pancreas. It flows into the duodenum and contains amylase for digesting starch, lipase for digesting fats and protease for digesting proteins.

Intestinal juice

Intestinal juice is produced by glands in the lining of the small intestine. It contains enzymes for digesting disaccharides and proteins as well as mucus and salts to make the contents of the small intestine more alkaline so the enzymes can work.

Absorption

The small molecules produced by digestion are absorbed into the villi of the wall of the small intestine. The tiny finger-like projections of the villi increase the surface area for absorption. Glucose and amino acids pass directly through the wall into the blood stream by diffusion or active transport. Fatty acids and glycerol enter vessels of the lymphatic system (lacteals) that run up the centre of each villus.

The Liver

The liver is situated in the abdominal cavity adjacent to the diaphragm (2 and 14). It is the largest single organ of the body and has over 100 known functions. Its most important digestive functions are:

1. the production of bile to help the digestion of fats (described above) and
2. the control of blood sugar levels

Glucose is absorbed into the capillaries of the villi of the intestine. The blood stream takes it directly to the liver via a blood vessel known as the hepatic portal vessel or vein.

The liver converts this glucose into glycogen which it stores. When glucose levels are low the liver can convert the glycogen back into glucose. It releases this back into the blood to keep the level of glucose constant. The hormone insulin, produced by special cells in the pancreas, controls this process.

Homeostasis

The cells of an animal can only remain healthy if the conditions are just right. The processes that take place in them are upset if the temperature is too high or too low, or if the fluid around or inside them is too acid or alkaline. Homeostasis is the name given to the processes that help keep the internal conditions constant even when external conditions change. The word means, "staying the same".

There are a number of organs in the body that play a part in maintaining homeostasis. For example, the skin helps keep the internal temperature of bird and mammals bodies within a narrow range even when the outside temperatures change; the lungs control the amount of carbon dioxide in the blood; the liver and pancreas work together to keep the amount of glucose in the blood within narrow limits and the kidneys regulate the acidity and the concentration of water and salt in the blood. How the kidneys do this will be described later in this chapter.

Hormones are chemicals that carry messages around the body in the blood and are central to many of the homeostatic processes mentioned above.

WATER IN THE BODY

Water is essential for living things to survive because all the chemical reactions within a body take place in a solution of water. An animal's body consists of up to 80% water. The exact proportion depends on the type of animal, its age, sex, health and whether or not it has had sufficient to drink. Generally animals do not survive a loss of more than 15% of their body water.

In vertebrates almost 2/3rd of this water is in the cells (intracellular fluid). The rest is outside the cells (extracellular fluid) where it is found in the spaces around the cells (tissue fluid), as well as in the blood and lymph.

Maintaining Water Balance

Animals lose water through their skin and lungs, in the faeces and urine. These losses must be made up by water in food and drink and from the water that is a by-product of chemical reactions. If the animal does not manage to compensate for water loss the dissolved substances in the blood may become so concentrated they become lethal. To prevent this happening various mechanisms come into play as soon as the concentration of the blood increases. A part of the brain called the hypothalamus is in charge of these homeostatic processes. The most important is the feeling of thirst that is triggered by an increase in blood concentration. This stimulates an animal to find water and drink it.

The kidneys are also involved in maintaining water balance as various hormones instruct them to produce more concentrated urine and so retain some of the water that would otherwise be lost.

Desert Animals

Coping with water loss is a particular problem for animals that live in dry conditions. Some, like the camel, have developed great tolerance for dehydration. For example, under some conditions, camels can withstand the loss of one third of their body mass as water. They can also survive wide daily changes in temperature. This means they do not have to use large quantities of water in sweat to cool the body by evaporation.

Smaller animals are more able than large ones to avoid extremes of temperature or dry conditions by resting in sheltered more humid situations during the day and being active only at night.

The kangaroo rat is able to survive without access to any drinking water at all because it does not sweat and produces extremely concentrated urine. Water from its food and from chemical processes is sufficient to supply all its requirements.

Excretion

Animals need to excrete because they take in substances that are excess to the body's requirements and many of the chemical reactions in the body produce waste products. If these substances were not removed they would poison cells or slow down metabolism. All animals therefore have some means of getting rid of these wastes.

The major waste products in mammals are carbon dioxide that is removed by the lungs, and urea that is produced when excess amino acids (from proteins) are broken down. Urea is filtered from the blood by the kidneys.

THE KIDNEYS AND URINARY SYSTEM

The kidneys in mammals are bean-shaped organs that lie in the abdominal cavity attached to the dorsal wall on either side of the spine. An artery from the dorsal aorta called the renal artery supplies blood to them and the renal vein drains them.

To the naked eye kidneys seem simple enough organs. They are covered by a fibrous coat or capsule and if cut in half lengthways (longitudinally) two distinct regions can be seen - an inner region or medulla and the outer cortex. A cavity within the kidney called the pelvis collects the urine and carries it to the ureter, which connects with the bladder where the urine is stored temporarily. Rings of muscle (sphincters) control the release of urine from the bladder and the urine leaves the body through the urethra.

Kidney Tubules or Nephrons

It is only when you examine kidneys under the microscope that you find that their structure is not simple at all. The cortex and medulla are seen to be composed of masses of tiny tubes. These are called kidney tubules or nephrons. A human kidney consists of over a million of them.

At one end of each nephron, in the cortex of the kidney, is a cup shaped structure called the (Bowman's or renal) capsule. It surrounds a tuft of capillaries called the glomerulus that carries high-pressure blood. Together the glomerulus and capsule act as a blood-filtering device. The holes in the filter allow most of the contents of the blood through except the red and white cells and large protein molecules. The fluid flowing from the capsule into the rest of the kidney tubule is therefore very similar to blood plasma and contains many useful substances like water, glucose, salt and amino acids. It also contains waste products like urea.

Processes Occurring In The Nephron

After entering the glomerulus the filtered fluid flows along a coiled part of the tubule (the proximal convoluted tubule) to a looped portion (the Loop of Henle) and then to the collecting tube via a second length of coiled tube (the distal convoluted tubule). From the collecting ducts the urine flows into the renal pelvis and enters the ureter.

Note that the glomerulus, capsule and both coiled parts of the tubule are all situated in the cortex of the kidney while the loops of Henle and collecting ducts make up the medulla.

As the fluid flows along the proximal convoluted tubule useful substances like glucose, water, salts, potassium ions, calcium ions and amino acids are reabsorbed into the blood capillaries that form a network around the tubules. Many of these substances are transported by active transport and energy is required.

In a separate process, some substances, particularly potassium, ammonium and hydrogen ions, and drugs like penicillin, are actively secreted into the distal convoluted tubule.

By the time the fluid has reached the collecting ducts these processes of absorption and secretion have changed the fluid originally filtered into the Bowman's capsule into urine. The main function of the collecting ducts is then to remove more water from the urine if necessary.

Normal urine consists of water, in which waste products such as urea and salts such as sodium chloride are dissolved. Pigments from the breakdown of red blood cells give urine its yellow colour.

The Production Of Concentrated Urine

Because of the high pressure of the blood in the glomerulus and the large size of the pores in the glomerulus/capsule-filtering device, an enormous volume of fluid passes into the kidney tubules. If this fluid were left as it is, the animal's body would be drained dry in 30 minutes. In fact, as the fluid flows down the tubule, over 90% of the water in it is reabsorbed. The main part of this reabsorption takes place in the collecting tubes.

The amount of water removed from the collecting ducts is controlled by a hormone called antidiuretic hormone (ADH) produced by the pituitary gland, situated at the base of the brain. When the blood becomes more concentrated, as happens when an animal is deprived of water, ADH is secreted and causes more water to be absorbed from the collecting ducts so that concentrated urine is produced. When the animal has drunk plenty of water and the blood is dilute, no ADH is secreted and no or little water is absorbed from the collecting ducts, so dilute urine is produced. In this way the concentration of the blood is controlled precisely.

WATER BALANCE IN FISH AND MARINE ANIMALS

Fresh Water Fish

Although the skin of fish is more or less waterproof, the gills are very porous. The body fluids of fish that live in fresh water have a higher concentration of dissolved substances than the water in which they swim. In other words the body fluids of fresh water fish are hypertonic to the water. Water therefore flows into the body by osmosis. To stop the body fluids being constantly diluted fresh water fish produce large quantities of dilute urine.

Marine Fish

Marine fish like the sharks and dogfish have body fluids that have the same concentration of dissolved substances as the water (isotonic) have little problem with water balance. However, marine bony fish like red cod, snapper and sole, have body fluids with a lower concentration of dissolved substances than seawater (they are hypotonic to seawater). This means that water tends to flow out of their bodies by osmosis. To make up this fluid loss they drink seawater and get rid of the excess salt by excreting it from the gills.

Marine Birds

Marine birds that eat marine fish take in large quantities of salt and some only have access to seawater for drinking. Bird's kidneys are unable to produce very concentrated urine, so they have developed a salt gland. This excretes a concentrated salt solution into the nose to get rid of the excess salt.

DIABETES AND THE KIDNEY

There are two types of diabetes. The most common is called sugar diabetes or diabetes mellitus and is common in cats and dogs especially if they are overweight. It is caused by the pancreas secreting insufficient insulin, the hormone that controls the amount of glucose in the blood. If insulin secretion is inadequate, the concentration of glucose in the blood increases. Any increase in the glucose in the blood automatically leads to an increase in glucose in the fluid filtered into the kidney tubule. Normally the kidney removes all the glucose filtered into it, but these high concentrations swamp this removal mechanism and urine containing glucose is produced. The main symptoms of this type of diabetes are the production of large amounts of dilute urine containing glucose, and excessive thirst.

The second type of diabetes is called diabetes insipidus. The name comes from the main symptom, which is the production of large amounts of very dilute and "tasteless" urine. It occurs when the pituitary gland produces insufficient ADH, the hormone that stimulates water re-absorption from the kidney tubule. When this hormone is lacking, water is not absorbed and large amounts of dilute urine are produced. Because so much water is lost in the urine, animals with this form of diabetes can die if deprived of water for only a day or so.

Other Functions Of The Kidney

The excretion of urea from the body and the maintenance of water balance, as described above, are the main functions of the kidney. However, the kidneys have other roles in keeping conditions in the body stable i.e. in maintaining homeostasis. These include:

- controlling the concentration of salt ions ($Na+$, $K+$, $Cl-$) in the blood by adjusting how much is excreted or retained;
- maintaining the correct acidity of the blood. Excess acid is constantly being produced by the normal chemical reactions in the body and the kidney eliminates this.

NORMAL URINE

Normal urine consists of water (95%), urea, salts (mostly sodium chloride) and pigments (mostly from bile) that give it its characteristic colour.

Abnormal Ingredients Of Urine

If the body is not working properly, small amounts of substances not normally present may be found in the urine or substances normally present may appear in abnormal amounts.

- The presence of glucose may indicate diabetes (see above).
- Urine with red blood cells in it is called haematuria, and may indicate inflammation of the kidney,or urinary tract, cancer or a blow to the kidneys.
- Sometimes free haemoglobin is found in the urine. This indicates that the red blood cells in the blood have haemolysed (the membrane has broken down) and the haemoglobin has passed into the kidney tubules.
- The presence of white blood cells in the urine indicates there is an infection in the kidney or urinary tract.
- Protein molecules are usually too large to pass into the kidney tubule so no or only small amounts of proteins like albumin is normally found in urine. Large quantities of albumin indicate that the kidney tubules have been injured or the kidney has become diseased. High blood pressure also pushes proteins from the blood into the tubules.
- Casts are tiny cylinders of material that have been shed from the lining of the tubules and flushed out into the urine.
- Mucus is not usually found in the urine of healthy animals but is a normal constituent of horses' urine, giving it a characteristic cloudy appearance.

Tests can be carried out to identify any abnormal ingredients of urine. These tests are normally done by "stix", which are small plastic strips with absorbent ends impregnated with various chemicals. A colour change occurs in the presence of an abnormal ingredient.

Excretion In Birds

Birds' high body temperature and level of activity means that they need to conserve water. Birds therefore do not have a bladder and instead of excreting urea, which needs to be dissolved in large amounts of water, birds produce uric acid that can be discharged as a thick paste along with the feces. This is the white chalky part of the bird droppings that land on you or your car.

Reproductive System

In biological terms sexual reproduction involves the union of gametes - the sperm and the ovum - produced by two parents. Each gamete is formed by meiosis (3). This means each contains only half the chromosomes of the body cells (haploid). Fertilization results in the joining of the male and female gametes to form a zygote which contains the full number of chromosomes (diploid). The zygote then starts to divide by mitosis (3) to form a new animal with all its body cells containing chromosomes that are identical to those of the original zygote.

The offspring formed by sexual reproduction contain genes from both parents and show considerable variation. For example, kittens in a litter are all different although they (usually) have the same mother and father. In the wild this variation is important because it means that when the environment changes some individuals may be better adapted to survive than others. These survivors pass their "superior" genes on to their offspring. In this way the characteristics of a group of animals can gradually change over time to keep pace with the changing environment. This "survival of the fittest" or "natural selection" is the mechanism behind the theory of evolution.

Fertilisation

In most fish and amphibia (frogs and toads) fertilisation of the egg cells takes place outside the body. The female lays the eggs and then the male deposits his sperm on or at least near them.

In reptiles and birds, eggs are fertilized inside the body when the male deposits the sperm inside the egg duct of the female. The egg is then surrounded by a resistant shell, "laid" by the female and the embryo completes its development inside the egg.

In mammals the sperm are placed in the body of the female and the eggs are fertilized internally. They then develop to quite an advanced stage inside the body of the female. When they are born they are fed on milk excreted from the mammary glands and protected by their parents until they become independent.

Sexual Reproduction In Mammals

The reproductive organs of mammals produce the gametes (sperm and egg cells), help them fertilise and then support the developing embryo.

The Male Reproductive System

The male reproductive system consists of a pair of testes that produce sperm (or spermatozoa), ducts that transport the sperm to the penis and glands that add secretions to the sperm to make semen.

The Testes

Sperm need temperatures between 2 to 10 degrees Centigrade lower and then the body temperature to develop. This is the reason why the testes are located in a bag of skin called the scrotal sacs (or scrotum) that hangs below the body and where the evaporation of secretions from special glands can further reduce the temperature. In many animals (including humans) the testes descend into the scrotal sacs at birth but in some animals they do not descend until sexual maturity and in others they only descend temporarily during the breeding season. A mature animal in which one or both testes have not descended is called a cryptorchid and is usually infertile.

The problem of keeping sperm at a low enough temperature is even greater in birds that have a higher body temperature than mammals. For this reason bird's sperm are usually produced at night when the body temperature is lower and the sperm themselves are more resistant to heat.

The testes consist of a mass of coiled tubes (the seminiferous or sperm producing tubules) in which the sperm are formed by meiosis. Cells lying between the seminiferous tubules produce the male sex hormone testosterone.

When the sperm are mature they accumulate in the collecting ducts and then pass to the epididymisbefore moving to the sperm duct or vas deferens. The two sperm ducts join the urethra just below the bladder, which passes through the penis and transports both sperm and urine.

Ejaculation discharges the semen from the erect penis. It is brought about by the contraction of the epididymis, vas deferens, prostate gland and urethra.

Semen

Semen consists of 10% sperm and 90% fluid and as sperm pass down the ducts from testis to penis, (accessory) glands add various secretion...

Accessory Glands

Three different glands may be involved in producing the secretions in which sperm are suspended, although the number and type of glands varies from species to species.

Seminal vesicles are important in rats, bulls, boars and stallions but are absent in cats and dogs. When present they produce secretions that make up much of the volume of the semen, and transport and provide nutrients for the sperm.

The prostate gland is important in dogs and humans. It produces an alkaline secretion that neutralizes the acidity of the male urethra and female vagina.

Cowper's glands (bulbourethral glands) have various functions in different species. The secretions may lubricate, flush out urine or form a gelatinous plug that traps the semen in the female reproductive system after copulation and prevents other males of the same species fertilizing an already mated female. Cowper's glands are absent in bears and aquatic mammals.

The Penis

The penis consists of connective tissue with numerous small blood spaces in it. These fill with blood during sexual excitement causing erection.

Penis Form And Shape

Dogs, bears, seals, bats and rodents have a special bone in the penis which helps maintain the erection. In some animals (e.g. the bull, ram and boar) the penis has an "S" shaped bend that allows it to fold up when not in use. In many animals the shape of the penis is adapted to match that of the vagina. For example, the boar has a corkscrew shaped penis, there is a pronounced twist in bulls' and it is forked in marsupials like the opossum. Some have spines, warts or hooks on them to help keep them in the vagina and copulation may be extended to help retain the semen in the female system. Mating can last up to three hours in minks, and dogs may "knot" or "tie" during mating and can not separate until the erection has subsided.

Sperm

Sperm are made up of three parts: a head consisting mainly of the nucleus, a midpiece containing many mitochondria to provide the energy and a tail that provides propulsion.

A single ejaculation may contain 2-3 hundred million sperm but even in normal semen as many as 10% of these sperm may be abnormal and infertile. Some may be dead while others are inactive or deformed with double, giant or small heads or tails that are coiled or absent altogether.

When there are too many abnormal sperm or when the sperm concentration is low, the semen may not be able to fertilize an egg and the animal is infertile. Make sure you don't confuse infertility with impotence, which is the inability to copulate successfully.

Sperm do not live forever. They have a definite life span that varies from species to species. They survive for between 20 days (guinea pig) to 60 days (bull) in the epididymis but once ejaculated into the female tract they only live from 12 to 48 hours. When semen is used for artificial insemination, storage under the right conditions can extend the life span of some species.

ARTIFICIAL INSEMINATION

In many species the male can be artificially stimulated to ejaculate and the semen collected. It can then be diluted, stored and used to inseminate females. For example bull semen can be diluted and stored for up to 3 weeks at room temperature. If mixed with an antifreeze solution and stored in "straws" in liquid nitrogen at minus 79C it will keep for much longer. Unfortunately the semen of chickens, stallions and boars can only be stored for up to 2 days.

Dilution of the semen means that one male can be used to fertilise many more females than would occur under natural conditions. There are also advantages in the male and female not having to make physical contact. It means that owners of females do not have to buy expensive males and the possibility of transmitting sexually transmitted diseases is reduced. Routine examination of the semen for sperm concentration, quality and activity allows only the highest quality semen to be used so a high success rate is ensured.

Since the lifespan of sperm in the female tract is so short and ova only survive from 8 to 10 hours the timing of the artificial insemination is critical. Successful conception depends upon detecting the time that the animal is "on heat" and when ovulation occurs.

The Female Reproductive Organs

The female reproductive system consists of a pair of ovaries that produce egg cells or ova and fallopian tubes where fertilisation occurs and which carry the fertilised ovum to the uterus. Growth of the foetus takes place here. The cervix separates the uterus from the vagina or birth canal, where the sperm are deposited.

Note that primates like humans have a uterus with a single compartment but in most mammals the uterus is divided into two separate parts or horns.

The Ovaries

Ovaries are small oval organs situated in the abdominal cavity just ventral to the kidneys. Most animals have a pair of ovaries but in birds only the left one is functional to reduce weight.

The ovary consists of an inner region (medulla) and an outer region (cortex) containing egg cells or ova. These are formed in large numbers around the time of birth and start to develop after the animal becomes sexually mature. A cluster of cells called the follicle surrounds and nourishes each ovum.

THE OVARIAN CYCLE

The ovarian cycle refers to the series of changes in the ovary during which the follicle matures, the ovum is shed and the corpus luteum develops (13.7).

Numerous undeveloped ovarian follicles are present at birth but they start to mature after sexual maturity. In animals that normally have only one baby at a time only one ovum will mature at once but in litter animals several will. The mature follicle consists of outer cells that provide nourishment. Inside this is a fluid-filled space that contains the ovum.

A mature follicle can be quite large, ranging from a few millimetres in small mammals to the size of a golf ball in large animals. It bulges out from the surface of the ovary before eventually rupturing to release the ovum into the abdominal cavity. Once the ovum has been shed, a blood clot forms in the empty follicle. This develops into a tissue called the corpus luteum that produces the hormone progesterone (13.9). If the animal becomes pregnant the corpus luteum persists, but if there is no pregnancy it degenerates and a new ovarian cycle usually.

The ovarian cycle showing from the top left clockwise: the maturation of the ovum over time, followed by ovulation and the development of the corpus luteum in the empty follicle

The Ovum

When the ovum is shed the nucleus is in the final stages of meiosis (cell division). It is surrounded by few layers of follicle cells and a tough membrane called the zona pellucida.

The Oestrous Cycle

The oestrous cycle is the sequence of hormonal changes that occurs through the ovarian cycle. These changes influence the behaviour and body changes of the female.

The first hormone involved in the oestrous cycle is follicle stimulating hormone (F.S.H.), secreted by the anterior pituitary gland. It stimulates the follicle to develop. As the follicle matures the outer cells begin to secrete the hormone oestrogen and this stimulates the mammary glands to develop. It also prepares the lining of the uterus to receive a fertilised egg. Ovulation is initiated by a surge of another hormone from the anterior pituitary, luteinising hormone (L.H.). This hormone also influences the development of the corpus luteum, which produces progesterone, a hormone that prepares the lining of the uterus for the fertilised ovum and readies the mammary glands for milk production. If no pregnancy takes place the corpus luteum shrinks and the production of progesterone decreases. This causes FSH to be produced again and a new oestrous cycle begins.

For fertilisation of the ovum by the sperm to occur, the female must be receptive to the male at around the time of ovulation. This is when the hormones turn on the signs of "heat", and she is "in season" or "in oestrous". These signs are turned off again at the end of the oestrous cycle.

During the oestrous cycle the lining of the uterus (endometrium) thickens ready for the fertilised ovum to be implanted. If no pregnancy occurs this thickened tissue is absorbed and the next cycle starts. In humans and other higher primates, however, the endometrium is shed as a flow of blood and instead of an oestrous cycle there is a menstrual cycle.

The length of the oestrous cycle varies from species to species. In rats the cycle only lasts 4-5 days and they are sexually receptive for about 14 hours. Dogs have a cycle that lasts 60-70 days and heat lasts 7-9 days and horses have a 21-day cycle and heat lasts an average of 6 days.

Ovulation is spontaneous in most animals but in some, e.g. the cat, and the rabbit, ovulation is stimulated by mating. This is called induced ovulation.

Signs of Oestrous or Heat

- When on heat a bitch has a blood stained discharge from the vulva that changes a little later to a straw coloured one that attracts all the dogs in the neighbourhood.
- Female cats "call" at night, roll and tread the carpet and are generally restless but will "stand" firm when pressure is placed on the pelvic region (this is the lordosis response).
- A female rat shows the lordosis response when on heat. It will "mount" other females and be more active than normal.
- A cow mounts other cows (bulling), bellows, is restless and has a discharge from the vulva.

Breeding Seasons and Breeding Cycles

Only a few animals breed throughout the year. This includes the higher primates (humans, gorillas and chimpanzees etc.), pigs, mice and rabbits. These are known as continuous breeders.

Most other animals restrict reproduction to one or two seasons in the year-seasonal breeders. There are several reasons for this. It means the young can be born at the time (usually spring) when feed is most abundant and temperatures are favourable. It is also sensible to restrict the breeding season because courtship, mating, gestation and the rearing of young can exhaust the energy resources of an animal as well as make them more vulnerable to predators.

The timing of the breeding cycle is often determined by day length. For example the shortening day length in autumn will bring sheep and cows into season so the foetus can gestate through the winter and be born in spring. In cats the increasing day length after the winter solstice (shortest day) stimulates breeding. The number of times an animal comes into season during the year varies, as does the number of oestrous cycles during each season. For example a dog usually has 2-3 seasons per year, each usually consisting of just one oestrous cycle. In contrast ewes usually restrict breeding to one season and can continue to cycle as many as 20 times if they fail to become pregnant.

FERTILISATION AND IMPLANTATION

Fertilisation

The opening of the fallopian tube lies close to the ovary and after ovulation the ovum is swept into its funnel-like opening and is moved along it by the action of cilia and wave-like contractions of the wall.

Copulation deposits several hundred million sperm in the vagina. They swim through the cervix and uterus to the fallopian tubes moved along by whip-like movements of their tails and contractions of the uterus. During this journey the sperm undergo their final phase of maturation so they are ready to fertilise the ovum by the time they reach it in the upper fallopian tube.

High mortality means only a small proportion of those deposited actually reach the ovum. The sperm attach to the outer zona pellucida and enzymes secreted from a gland in the head of the sperm dissolve this membrane so it can enter. Once one sperm has entered, changes in the zona pellucida prevent further sperm from penetrating. The sperm loses its tail and the two nuclei fuse to form a zygote with the full set of paired chromosomes restored.

Development Of The Morula And Blastocyst

As the fertilised egg travels down the fallopian tube it starts to divide by mitosis. First two cells are formed and then four, eight, sixteen, etc. until there is a solid ball of cells. This is called a morula. As division continues a hollow ball of cells develops. This is a blastocyst.

Implantation

Implantation involves the blastocyst attaching to, and in some species, completely sinking into the wall of the uterus.

PREGNANCY

The Placenta And Foetal Membranes

As the embryo increases in size, the placenta, umbilical cord and foetal membranes (often known collectively as the placenta) develop to provide it with nutrients and remove waste products. In later stages of development the embryo becomes known as a foetus.

The placenta is the organ that attaches the foetus to the wall of the uterus. In it the blood of the foetus and mother flow close to each other but never mix. The closeness of the maternal and foetal blood systems allows diffusion between them. Oxygen and nutrients diffuse from the mother's blood into that of the foetus and carbon dioxide and excretory products diffuse in the other direction. Most maternal hormones (except adrenaline), antibodies, almost all drugs (including alcohol), lead and DDT also pass across the placenta. However, it protects the foetus from infection with bacteria and most viruses.

The foetus is attached to the placenta by the umbilical cord. It contains arteries that carry blood to the placenta and a vein that returns blood to the foetus. The developing foetus becomes surrounded by membranes. These enclose the amniotic fluid that protects the foetus from knocks and other trauma.

Hormones During Pregnancy

The corpus luteum continues to secrete progesterone and oestrogen during pregnancy. These maintain the lining of the uterus and prepare the mammary glands for milk secretion. Later in the pregnancy the placenta itself takes over the secretion of these hormones.

Chorionic gonadotrophin is another hormone secreted by the placenta and placental membranes. It prevents uterine contractions before labour and prepares the mammary glands for lactation. Towards the end of pregnancy the placenta and ovaries secrete relaxin, a hormone that eases the joint between the two parts of the pelvis and helps dilate the cervix ready for birth.

Pregnancy Testing

The easiest method of pregnancy detection is ultrasound which is noninvasive and very reliable Later in gestation pregnancy can be detected by taking x-rays.

In dogs and cats a blood test can be used to detect the hormone relaxin.

In mares and cows palpation of the uterus via the rectum is the classic way to determine pregnancy. It can also be done by detecting the hormones progesterone or equine chorionic gonagotrophin (eCG) in the urine. A new sensitive test measures the amount of the hormone, oestrone sulphate, present in a sample of faeces. The hormone is produced by the foal and placenta, and is only present when there is a living foal.

In most animals, once pregnancy is advanced, there is a window of time during which an experienced veterinarian can determine pregnancy by feeling the abdomen.

Gestation Period

The young of many animals (e.g. pigs, horses and elephants) are born at an advanced state of development, able to stand and even run to escape predators soon after they are born. These animals have a relatively long gestation period that varies with their size e.g. from 114 days in the pig to 640 days in the elephant.

In contrast, cats, dogs, mice, rabbits and higher primates are relatively immature when born and totally dependent on their parents for survival. Their gestation period is shorter and varies from 25 days in the mouse to 31 days in rabbits and 258 days in the gorilla. The babies of marsupials are born at an extremely immature stage and migrate to the pouch where they attach to a teat to complete their development. Kangaroo joeys, for example, are born 33 days after conception and opossums after only 8 days.

BIRTH

Signs Of Imminent Birth

As the pregnancy continues, the mammary glands enlarge and may secrete a milky substance a few days before birth occurs. The vulva may swell and produce thick mucus and there is sometimes a visible change in the position of the foetus. Just before birth the mother often becomes restless, lying down and getting up frequently. Many animals seek a secluded place where they may build a nest in which to give birth.

Labour

Labour involves waves of uterine contractions that press the foetus against the cervix causing it to dilate. The foetus is then pushed through the cervix and along the vagina before being delivered. In the final stage of labour the placenta or "afterbirth" is expelled.

Adaptations Of The Foetus To Life Outside The Uterus

The foetus grows in the watery, protected environment of the uterus where the mother supplies oxygen and nutrients, and waste products pass to her blood circulation for excretion. Once the baby animal is born it must start to breathe for itself, digest food and excrete its own waste. To allow these functions to occur blood is re-routed to the lungs and the glands associated with the gut start to secrete. Note that newborn animals can not control their own body temperature. They need to be kept warm by the mother, littermates and insulating nest materials.

MILK PRODUCTION

Cows, manatees and primates have two mammary glands but animals like pigs that give birth to large litters may have as many as 12 pairs. Ducts from the gland lead to a nipple or teat and there may be a sinus where the milk collects before being suckled.

The hormones oestrogen and progesterone stimulate the mammary glands to develop and prolactin promotes the secretion of the milk. Oxytocin from the pituitary gland releases the milk when the baby suckles. The first milk is called colostrum. It is a rich in nutrients and contains protective antibodies from the mother. Milk contains fat, protein and milk sugar as well as vitamins and most minerals although it contains little iron. Its actual composition varies widely from species to species. For example whale's and seal's milk has twelve times more fat and four times more protein than cow's milk. Cow's milk has far less protein in it than cat's or dog's milk. This is why orphan kittens and puppies cannot be fed cow's milk.

Reproduction In Birds

Male birds have testes and sperm ducts and male swans, ducks, geese and ostriches have a penis. However, most birds make do with a small amount of erectile tissue known as a papilla. To reduce weight for flight most female birds only have one ovary - usually the left, which produces extremely yolky eggs. The eggs are fertilised in the upper part of the oviduct (equivalent to the fallopian tube and uterus of mammals) and as they pass down it albumin (the white of the egg), the membrane beneath the shell and the shell are laid down over the yolk. Finally the egg is covered in a layer of mucus to help the bird lay it.

Most birds lay their eggs in a nest and the hen sits on them until they hatch. Ducklings and chicks are relatively well developed when they hatch and able to forage for their own food. Most other nestlings need their parents to keep them warm, clean and fed. Young birds grow rapidly and have voracious appetites that may involve the parents making up to 1000 trips a day to supply their need for food.

Nervous System

Animals must be able to sense and respond to the environment in which they live if they are to survive. They need to be able to sense the temperature of their surroundings, for example, so they can avoid the hot sun. They must also be able to identify food and escape predators.

The various systems and organs in the body must also be linked so they work together. For example, once a predator has identified suitable prey it has to catch it. This involves coordinating the contraction of the muscle so the predator can run, there must then be an increased blood supply to the muscles to provide them with oxygen and nutrients. At the same time the respiration rate must increase to supply the oxygen and remove the carbon dioxide produced as a result of this increased activity. Once the prey has been caught and eaten, the digestive system must be activated to digest it.

The adjustment of an animal's response to changes in the environment and the complex linking of the various processes in the body that this response involves are called co-ordination. Two systems are involved in co-ordination in animals. These are the nervous and endocrine systems. The first operates via electrical impulses along nerve fibres and the second by releasing special chemicals or hormones into the bloodstream from glands.

Functions of the Nervous System

The nervous system has three basic functions:

1. Sensory function - to sense changes (known as stimuli) both outside and within the body. For example the eyes sense changes in light and the ear responds to sound waves. Inside the body, stretch receptors in the stomach indicate when it is full and chemical receptors in the blood vessels monitor the acidity of the blood.
2. Integrative function - processing the information received from the sense organs. The impulses from these organs are analysed and stored as memory. The many different impulses from different sources are sorted, synchronised

and co-ordinated and the appropriate response initiated. The power to integrate, remember and apply experience gives higher animals much of their superiority.

3. Motor function - The third function is the response to the stimuli that causes muscles to contract or glands to secrete.

All nervous tissue is made up of nerve cells or neurons. These transmit high-speed signals called nerve impulses. Nerve impulses can be thought of as being similar to an electric current.

The Neuron

Neurons are cells that have been adapted to carry nerve impulses. A typical neuron has a cell body containing a nucleus, one or more branching filaments called dendrites which conduct nerve impulses towards the cell body and one long fibre, an axon, that carries the impulses away from it. Many axons have a sheath of fatty material called myelin surrounding them. This speeds up the rate at which the nerve impulses travel along the nerve.

The cell body of neurons is usually located in the brain or spinal cord while the axon extends the whole distance to the organ that it supplies. The neuron carrying impulses from the spinal cord to the hind leg or tail of a horse, for example, can be several feet long. A nerve is a bundle of axons.

A sensory neuron is a nerve cell that transmits impulses from a sense receptor such as those in the eye or ear to the brain or spinal cord. A motor neuron is a nerve cell that transmits impulses from the brain or spinal cord to a muscle or gland. A relay neuron connects sensory and motor neurons and is found in the brain or spinal cord.

CONNECTIONS BETWEEN NEURONS

The connection between adjacent neurons is called a synapse. The two nerve cells do not actually touch here for there is a microscopic space between them. The electrical impulse in the neurone before the synapse stimulates the production of chemicals called neurotransmitters (such as acetylcholine), which are secreted into the gap.

The neurotransmitter chemicals diffuse across the gap and when they contact the membrane of the next nerve cell they stimulate a new nervous impulse. After the impulse has passed the chemical is destroyed and the synapse is ready to receive the next nerve impulse.

Reflexes

A reflex is a rapid automatic response to a stimulus. When you accidentally touch a hot object and automatically jerk your hand away, this is a reflex action. It happens without you having to think about it. Animals automatically blink when an object approaches the eye and cats twist their bodies in the air when falling so they land on their paws. (Please don't test this one at home with your pet cat!).

Swallowing, sneezing, and the constriction of the pupil of the eye in bright light are also all reflex actions.

The path taken by the nerve impulses in a reflex is called a reflex arc. Most reflex arcs involve only three neurons. The stimulus (a pin in the paw) stimulates the pain receptors of the skin, which initiate an impulse in a sensory neuron. This travels to the spinal cord where it passes, by means of a synapse, to a connecting neuron called the relay neuron situated in the spinal cord. The relay neuron in turn makes a synapse with one or more motor neurons that transmit the impulse to the muscles of the limb causing them to contract and remove the paw from the sharp object. Reflexes do not require involvement of the brain although you are aware of what is happening and can, in some instances, prevent them happening. Animals are born with their reflexes. You can think of them as being wired in.

Conditioned Reflexes

In most reflexes the stimulus and response are related. For example the presence of food in the mouth causes the salivary glands to release saliva. However, it is possible to train animals (and humans) to respond to different and often quite irrelevant stimuli. This is called a conditioned reflex.

A Russian biologist called Pavlov carried out the classic experiment to demonstrate such a reflex when he conditioned dogs to salivate at the sound of a bell ringing. Almost every pet owner can identify reflexes they have conditioned in their animals. Perhaps you have trained your cat to associate food with the opening of the fridge door or accustomed your dog to the routines you go through before taking them for a walk.

Parts of the Nervous System

When we describe the nervous system of vertebrates we usually divide it into two parts.

1. The central nervous system (CNS) which consists of the brain and spinal cord.
2. The peripheral nervous system (PNS) which consists of the nerves that connect to the brain and spinal cord (cranial and spinal nerves) as well as the autonomic (or involuntary) nervous system.

The Central Nervous System

The central nervous system consists of the brain and spinal cord. It acts as a kind of 'telephone exchange' where a vast number of cross connections are made.

When you look at the brain or spinal cord some regions appear creamy white (white matter) and others appear grey (grey matter). White matter consists of masses of nerve axons and the grey matter consists of the nerve cells. In the brain the grey matter is on the outside and in the spinal cord it is on the inside.

The Brain

The major part of the brain lies protected within the sturdy "box" of skull called the cranium. Surrounding the fragile brain tissue (and spinal cord) are protective membranes called the meninges and a crystal-clear fluid called cerebrospinal fluid, which protects and nourishes the brain tissue. This fluid also fills four cavities or ventricles that lie within the brain.

Brain tissue is extremely active and, even when an animal is resting, it uses up to 20% of the oxygen taken into the body by the lungs. The carotid artery, a branch off the dorsal aorta, supplies it with the oxygen and nutrients it requires. Brain damage occurs if brain tissue is deprived of oxygen for only 4-8 minutes.

The brain consists of three major regions:

1. the fore brain which includes the cerebral hemispheres, hypothalamus and pituitary gland;
2. the hind brain or brain stem, contains the medulla oblongata and pons and
3. the cerebellum or "little brain".

Mapping the brain

In humans and some animals the functions of the different regions of the cerebral cortex have been mapped .

The Forebrain

The cerebral hemispheres are the masses of brain tissue that sit on the top of the brain. The surface is folded into ridges and furrows called sulci (singular sulcus). They make this part of the brain look rather like a very large walnut kernel. The two hemispheres are separated by a deep groove although they are connected internally by a thick bundle of nerve fibres. The outer layer of each hemisphere is called the cerebral cortex and this is where the main functions of the cerebral hemispheres are carried out.

The cerebral cortex is large and convoluted in mammals compared to other vertebrates and largest of all in humans because this is where the so-called "higher centres" concerned with memory, learning, reasoning and intelligence are situated.

Nerves from the eyes, ears, nose and skin bring sensory impulses to the cortex where they are interpreted. Appropriate voluntary movements are initiated here in the light of the memories of past events.

Different regions of the cortex are responsible for particular sensory and motor functions, e.g. vision, hearing, taste, smell, or moving the fore-limbs, hind-limbs or tail. For example, when a dog sniffs a scent, sensory impulses from the organ of smell in the nose pass via the olfactory (smelling) nerve to the olfactory centres of the cerebral hemispheres where the impulses are interpreted and co-ordinated.

In humans and some animals the functions of the different regions of the cerebral cortex have been mapped.

The hypothalamus is situated at the base of the brain and is connected by a "stalk" to the pituitary gland, the "master" hormone-producing gland. The hypothalamus can be thought of as the bridge between the nervous and endocrine (hormone producing) systems. It produces some of the hormones that are released from the pituitary gland and controls the release of others from it.

It is also an important centre for controlling the internal environment of the animal and therefore maintaining homeostasis. For example, it helps regulate the movement of food through the gut and the temperature, blood pressure and concentration of the blood. It is also responsible for the feeling of being hungry or thirsty and it controls sleep patterns and sex drive.

THE HINDBRAIN

The medulla oblongata is at the base of the brain and is a continuation of the spinal cord. It carries all signals between the spinal cord and the brain and contains centres that control vital body functions like the basic rhythm of breathing, the rate of the heartbeat and the activities of the gut. The medulla oblongata also co-ordinates swallowing, vomiting, coughing and sneezing.

The Cerebellum

The cerebellum (little brain) looks rather like a smaller version of the cerebral hemispheres attached to the back of the brain. It receives impulses from the organ of balance (vestibular organ) in the inner ear and from stretch receptors in the muscles and tendons. By co-ordinating these it regulates muscle contraction during walking and running and helps maintain the posture and balance of the animal. When the cerebellum malfunctions it causes a tremor and uncoordinated movement.

The Spinal Cord

The spinal cord is a cable of nerve tissue that passes down the channel in the vertebrae from the hindbrain to the end of the tail. It becomes progressively smaller as paired spinal nerves pass out of the cord to parts of the body. Protective membranes or meninges cover the cord and these enclose cerebral spinal fluid.

The Peripheral Nervous System

The peripheral nervous system consists of nerves that are connected to the brain (cranial nerves), and nerves that are connected to the spinal cord (spinal nerves). The autonomic nervous system is also part of the peripheral nervous system.

CRANIAL NERVES

There are twelve pairs of cranial nerves that come from the brain. Each passes through a hole in the cranium (brain case). The most important of these are the olfactory, optic, acoustic and vagus nerves.

The olfactory nerves - (smell) carry impulses from the olfactory organ of the nose to the brain.

The optic nerves - (sight) carry impulses from the retina of the eye to the brain.

The auditory (acoustic) nerves - (hearing) carry impulses from the cochlear of the inner ear to the brain.

The vagus nerve - controls the muscles that bring about swallowing. It also controls the muscles of the heart, airways, lungs, stomach and intestines.

Spinal Nerves

Spinal nerves connect the spinal cord to sense organs, muscles and glands in the body. Pairs of spinal nerves leave the spinal cord and emerge between each pair of adjacent vertebrae.

The sciatic nerve is the largest spinal nerve in the body. It leaves the spinal cord as several nerves that join to form a flat band of nervous tissue. It passes down the thigh towards the hind leg where it gives off branches to the various muscles of this limb.

The Autonomic Nervous System

The autonomic nervous system controls internal body functions that are not under conscious control. For example when a prey animal is chased by a predator the autonomic nervous system automatically increases the rate of breathing and the heartbeat. It dilates the blood vessels that carry blood to the muscles, releases glucose from the liver, and makes other adjustments to provide for the sudden increase in activity. When the animal has escaped and is safe once again the nervous system slows down all these processes and resumes all the normal body activities like the digestion of food.

The nerves of the autonomic nervous system originate in the spinal cord and pass out between the vertebrae to serve the various organs.There are two main parts to the autonomic nervous system -- the sympathetic system and the parasympathetic system.

The sympathetic system stimulates the "flight, fright, fight" response that allows an animal to face up to an attacker or make a rapid departure. It increases the heart and respiratory rates, as well as the amount of blood flowing to the skeletal muscles while blood flow to less critical regions like the gut and skin is reduced. It also causes the pupils of the eyes to dilate. Note that the effects of the sympathetic system are similar to the effects of the hormone adrenaline.

The parasympathetic system does the opposite to the sympathetic system. It maintains the normal functions of the relaxed body. These are sometimes known as the "housekeeping" functions. It promotes effective digestion, stimulates defaecation and urination and maintains a regular heartbeat and rate of breathing.

The Senses

Sense organs allow animals to sense changes in the environment around them and in their bodies so that they can respond appropriately. They enable animals to avoid hostile environments, sense the presence of predators and find food.

Animals can sense a wide range of stimuli that includes, touch, pressure, pain, temperature, chemicals, light, sound, movement and position of the body. Some animals can sense electric and magnetic fields. All sense organs respond to stimuli by producing nerve impulses that travel to the brain via a sensory nerve. The impulses are then processed and interpreted in the brain as pain, sight, sound, taste etc.

The senses are often divided into two groups:

1. The general senses of touch, pressure, pain and temperature that are distributed fairly evenly through the skin. Some are found in muscles and within joints.
2. The special senses which include the senses of smell, taste, sight, hearing and balance. The special sense organs may be quite complex in structure.

Touch And Pressure

Within the dermis of the skin are numerous modified nerve endings that are sensitive to touch and pressure. The roots of hairs may also be well supplied with sensory receptors that inform the animal that it is in contact with an object. Whiskers are specially modified hairs.

Pain

Receptors that sense pain are found in almost every tissue of the body. They tell the animal that tissues are dangerously hot, cold, compressed or stretched or that there is not enough blood flowing in them. The animal may then be able to respond and protect itself from further damage

Temperature

Nerve endings in the skin respond to hot and cold stimuli.

Awareness Of Limb Position

There are sense organs in the muscles, tendons and joints that send continuous impulses to the brain that tell it where each limb is. This information allows the animal to place its limbs accurately and know their exact position without having to watch them.

Smell

Animals use the sense of smell to locate food, mark territory, identify their own offspring and the presence and sexual condition of a potential mate. The organ of smell (olfactory organ) is located in the nose and responds to chemicals in the air. It consists of modified nerve cells that have several tiny hairs on the surface. These emerge from the epithelium on the roof of the nose cavity into the mucus that lines it. As the animal breathes, chemicals in the air dissolve in the mucus. When the sense cell responds to a particular molecule, it fires an impulse that travels along the olfactory nerve to the brain where it is interpreted as an odour.

The olfactory sense in humans is rudimentary compared to that of many animals. Carnivores that hunt have a very highly developed sensitivity to scents. For example a polar bear can smell out a dead seal 20 km away and a bloodhound can distinguish between the trails of different people although it may sometimes be confused by the criss-crossing trail of identical twins.

Snakes and lizards detect odours by means of Jacobson's organ. This is situated on the roof of the mouth and consists of pits containing sensory cells. When snakes flick out their forked tongues they are smelling the air by carrying the molecules in it to the Jacobson's organ.

Taste

The sense of taste allows animals to detect and identify dissolved chemicals. In reptiles, birds, and mammals the taste receptors (taste buds) are found mainly to the upper surface of the tongue. They consist of pits containing sensory cells arranged rather like the segments of an orange. Each receptor cell has a tiny "hair" that projects into the saliva to sense the chemicals dissolved in it.

The sense of taste is quite restricted. Humans can only distinguish four different tastes (sweet, sour, bitter and salt) and what we normally think of as taste is mainly the sense of smell. Food is quite tasteless when the nose is blocked and cats often refuse to eat when this happens.

Sight

The eyes are the organs of sight. They consist of spherical eyeballs situated in deep depressions in the skull called the orbits. They are attached to the wall of the orbit by six muscles, which move the eyeball. Upper and lower eyelids cover the eyes during sleep and protect them from foreign objects or too much light, and spread the tears over their surface.

The nictitating membrane or hawis a transparent sheet that moves sideways across the eye from the inner corner, cleansing and moistening the cornea without shutting out the light. It is found in birds, crocodiles, frogs and fish as well as marsupials like the kangaroo. It is rare in mammals but can be seen in cats and dogs by gently opening the eye when it is asleep. Eyelashes also protect the eyes from the sun and foreign objects.

STRUCTURE OF THE EYE

Lining the eyelids and covering the front of the eyeball is a thin epithelium called the conjunctiva. Conjunctivitis is inflammation of this membrane. Tear glands that open just under the top eyelid secrete a salty solution that keeps the exposed part of the eye moist, washes away dust and contains an enzyme that destroys bacteria.

The wall of the eyeball is composed of three layers. From the outside these are the sclera, the choroid and the retina.

The sclera is a tough fibrous layer that protects the eyeball and gives it rigidity. At the front of the eye the sclera is visible as the "white" of the eye, which is modified as the transparent cornea through which the light rays have to pass to enter the eye. The cornea helps focus light that enters the eye.

The choroid lies beneath the sclera. It contains a network of blood vessels that supply the eye with oxygen and nutrients. Its inner surface is highly pigmented and absorbs stray light rays. In nocturnal animals like the cat and possum this highly pigmented layer reflects light as a means of conserving light. This is what makes them shine when caught in car headlights.

At the front of the eye the choroid becomes the iris. This is the coloured part of the eye that controls the amount of light entering the pupil of the eye. In dim light the pupil is wide open so as much light as possible enters while in bright light the pupils contract to protect the retina from damage by excess light.

The pupil in most animals is circular but in many nocturnal animals it is a slit that can close completely. This helps protect the extra-sensitive light sensing tissues of animals like the cat and possum from bright sunlight.

The inner layer lining the inside of the eye is the retina. This contains the light sensing cells called rods and cones.

The rod cells are long and fat and are sensitive to dim light but cannot detect colour. They contain large amounts of a pigment that changes when exposed to light. This pigment comes from vitamin A found in carrots etc. A deficiency of this vitamin causes night blindness. So your mother was right when she told you to eat your carrots as they would help you see in the dark!

The cone cells provide colour vision and allow animals to see details. Most are found in the centre of the retina and they are most densely concentrated in a small area called the fovea. This is the area of sharpest vision, where the words you are reading at this moment are focussed on your retina.

The nerve fibres from the cells of the retina join and leave the eye via the optic nerve. There are no rods or cones here and it is a blind spot. The optic nerve passes through the back of the orbit and enters the brain.

The lens is situated just behind the pupil and the iris. It is a crystalline structure with no blood vessels and is held in position by a ligament. This is attached to a muscle, which changes the shape of the lens so both near and distant objects can be focussed by the eye. This ability to change the focus of the lens is called accommodation. In many mammals the muscles that bring about accommodation are poorly developed, Rats, cows and dogs, for example, are thought to be unable to focus clearly on near objects.

In old age and certain diseases the lens may become cloudy resulting in blurred vision. This is called a cataract. Within the eyeball are two cavities, the anterior and posterior chambers, separated by the lens. They contain fluids the aqueous and vitreous humoursrespectively, that maintain the shape of the eyeball and help press the retina firmly against the choroid so clear images are seen.

How The Eye Sees

Eyes work quite like a camera. Light rays from an object enter the eye and are focused on the retina (the "film") at the back of the eye. The cornea, the lens and the fluid within the eye all help to focus the light. They do this by bending the light rays so that light from the object falls on the retina. This bending of light is called refraction. The light stimulates the light sensitive cells of the retina and nerve impulses are produced that pass down the optic nerve to the brain.

COLOUR VISION IN ANIMALS

As mentioned before, the retina has two different kinds of cells that are stimulated by light - 'rods' and 'cones'. In humans and higher primates like baboons and gorillas the rods function in dim light and do not perceive colour, while the cones are stimulated by bright light and perceive details and colour.

Other mammals have very few cones in their retinas and it is believed that they see no or only a limited range of colour. It is, of course, difficult to find out exactly what animals do see. It is thought that deer, rats, and rabbits and nocturnal animals like the cat are colour-blind, and dogs probably see green and blue. Some fish and most birds seem to have better colour vision than humans and they use colour, often very vivid ones, for recognizing each other as well as for courtship and protection.

Binocular Vision

Animals like cats that hunt have eyes placed on the front of the head in such a way that both eyes see the same wide area but from slightly different angles. This is called binocular vision. Its main advantage is that it enables the animals to estimate the distance to the prey so they can chase it and pounce accurately.

In contrast plant-eating prey animals like the rabbit and deer need to have a wide panoramic view so they can see predators approaching. They therefore have eyes placed on the side of the head, each with its own field of vision. They have only a very small area of binocular vision in front of the head but are extremely sensitive to movement.

Hearing

Animals use the sense of hearing for many different purposes. It is used to sense danger and enemies, to detect prey, to identify prospective mates and to communicate within social groups. Some animals (e.g. most bats and dolphins) use sound to "see" by echolocation. By sending out a cry and interpreting the echo, they sense obstacles or potential prey.

Structure of the Ear

Most of the ear, the organ of hearing, is hidden from view within the boney skull. It consists of three main regions: the outer ear, the middle ear and the inner ear.

The outer ear consists of an ear canal leading inwards to a thin membrane known as the eardrumor tympanic membrane that stretches across the canal. Many animals have an external ear flap or pinna to collect and funnel the sound into the ear canal. The pinnae (plural of pinna) usually face forwards on the head but many animals can swivel them towards the source of the sound.

In dogs the ear canal is long and bent and often traps wax or provides an ideal habitat for mites, yeast and bacteria.

The middle ear consists of a cavity in the skull that is connected to the pharynx (throat) by a long narrow tube called the Eustachian tube. This links the middle ear to the outside air so that the air pressure on both sides of the eardrum can be kept the same. Everyone knows the uncomfortable feeling (and affected hearing) that occurs when you drive down a steep hill and the unequal air pressures on the two sides of the eardrum cause it to distort. The discomfort is relieved when you swallow because the Eustachian tubes open and the pressure on either side equalises.

Within the cavity of the middle ear are three of the smallest bones in the body, the auditory ossicles. They are known as the hammer, the anvil and the stirrup because of their resemblance to the shape of these objects. These tiny bones articulate (move against) each other and transfer the vibrations of the eardrum to the membrane covering the opening to the inner ear.

The inner ear is a complicated series of fluid-filled tubes imbedded in the bone of the skull. It consists of two main parts. These are the cochlea where sound waves are converted to nerve impulses and the vestibular organ that is associated with the sense of balance and has no role in hearing.

The cochlea looks rather like a coiled up snail shell. Within it there are specialised cells with fine hairs on their surface that respond to the movement of the fluid within the cochlea by producing nervous impulses that travel to the brain along the auditory nerve.

How The Ear Hears

Sound waves can be thought of as vibrations in the air. They are collected by the ear pinna and pass down the ear canal where they cause the eardrum to vibrate. (An interesting fact is that when you are listening to someone speaking your eardrum vibrates at exactly the same rate as the vocal cords of the person speaking to you).

The vibration of the eardrum sets the three tiny bones in the middle ear moving against each other so that the vibration is transferred to the membrane covering the opening to the inner ear. As well as transferring the vibration, the tiny ear bones also amplify it. The three tiny bones are called the stirrup, anvil, and hammer. They were called such of their form. In the human ear this amplification is about 20 times while in desert-dwelling animals like the kangaroo rat it is 100 times. This acute hearing warns them of the approach of predators like owls and snakes, even in the dark.

The vibration causes waves in the fluid in the inner ear that pass down the cochlea. These waves stimulate the tiny hair cells to produce nerve impulses that travel via the auditory nerve to the cerebral cortex of the brain where they are interpreted as sound.

The hair cells generate a nerve impulse that travels down the auditory nerve to the brain.

Remember that sound waves do not pass along the Eustachian tube. Its function is to equalise the air pressure on either side of the tympanic membrane.

Balance

The vestibular organ of the inner ear helps an animal maintain its posture and keep balanced by monitoring the movement and position of the head. It consists of two structures - the semicircular canals and the otolith organs.

The semicircular canals respond to movement of the body. They tell an animal whether it is moving up or down, left or right. They consist of three canals set in three different planes at right angles to each other so that movement in any direction can be registered. The canals contain fluid and sense cells with fine hairs that project into the fluid. When the head moves the fluid swirls in the canals and stimulates the hair cells. These send nerve impulses along the vestibular nerve to the cerebellum.

Note that the semicircular canals register acceleration and deceleration as well as changes in direction but do not respond to movement that is at a constant speed.

The otolith organs are sometimes known as gravity receptors. They tell you if your head is tilted or if you are standing on your head. They consist of bulges at the base of the semi circular canals that contain hair cells that are covered by a mass of jelly containing tiny pieces of chalk called otoliths. When the head is tilted, or moved suddenly, the otoliths pull on the hair cells, which produce a nerve impulse. This travels down the vestibular nerve to the cerebellum. By coordinating the nerve impulses from the semicircular canals and otolith organs the cerebellum helps the animal keep its balance.

ENDOCRINE SYSTEM

In order to survive, animals must constantly adapt to changes in the environment. The nervous and endocrine systems both work together to bring about this adaptation. In general the nervous system responds rapidly to short-term changes by sending electrical impulses along nerves and the endocrine system brings about longer-term adaptations by sending out chemical messengers called hormones into the blood stream. For example, think about what happens when a male and female cat meet under your bedroom window at night. The initial response of both cats may include spitting, fighting and spine tingling yowling - all brought about by the nervous system. Fear and stress then activates the adrenal glands to secrete the hormone adrenaline which increases the heart and respiratory rates. If mating occurs, other hormones stimulate the release of ova from the ovary of the female and a range of different hormones maintains pregnancy, delivery of the kittens and lactation.

Endocrine Glands And Hormones

Hormones are chemicals that are secreted by endocrine glands. Unlike exocrine glands (5), endocrine glands have no ducts, but release their secretions directly into the blood system, which carries them throughout the body. However, hormones only affect the specific target organs that recognize them. For example, although it is carried to virtually every cell in the body, follicle stimulating hormone (FSH), released from the anterior pituitary gland, only acts on the follicle cells of the ovaries causing them to develop.

A nerve impulse travels rapidly and produces an almost instantaneous response but one that lasts only briefly. In contrast, hormones act more slowly and their effects may be long lasting. Target cells respond to minute quantities of hormones and the concentration in the blood is always extremely low. However, target cells are sensitive to subtle changes in hormone concentration and the endocrine system regulates processes by changing the rate of hormone secretion.

The main endocrine glands in the body are the pituitary, pineal, thyroid, parathyroid, and adrenal glands, the pancreas, ovaries and testes.

The Pituitary Gland And Hypothalamus

The pituitary gland is a pea-sized structure that is attached by a stalk to the underside of the cerebrum of the brain. It is often called the "master" endocrine gland because it controls many of the other endocrine glands in the body. However, we now know that the pituitary gland is itself controlled by the hypothalamus. This small but vital region of the brain lies just above the pituitary and provides the link between the nervous and endocrine systems. It controls the autonomic nervous system, produces a range of hormones and regulates the secretion of many others from the pituitary gland (7 for more information on the hypothalamus).

The pituitary gland is divided into two parts with different functions - the anterior and posterior pituitary.

The anterior pituitary gland secretes hormones that regulate a wide range of activities in the body. These include:

1. Growth hormone that stimulates body growth.
2. Prolactin that initiates milk production.
3 Follicle stimulating hormone (FSH) that stimulates the development of the follicles of the ovaries. These then secrete oestrogen (6).

The Pineal Gland

The pineal gland is found deep within the brain. It is sometimes known as the 'third eye" as it responds to light and day length. It produces the hormone melatonin, which influences the development of sexual maturity and the seasonality of breeding and hibernation.

The Thyroid Gland

The thyroid gland is situated in the neck, just in front of the windpipe or trachea. It produces the hormone thyroxine, which influences the rate of growth and development of young animals. In mature animals it increases the rate of chemical reactions in the body.

Thyroxine consists of 60% iodine and too little in the diet can cause goitre, an enlargement of the thyroid gland. Many inland soils in New Zealand contain almost no iodine so goitre can be common in stock when iodine supplements are not given. To add to the problem, chemicals called goitrogens that occur naturally in plants like kale that belong to the cabbage family, can also cause goitre even when there is adequate iodine available.

The Parathyroid Glands

The parathyroid glands are also found in the neck just behind the thyroid glands. They produce the hormone parathormone that regulates the amount of calcium in the blood and influences the excretion of phosphates in the urine.

The Adrenal Gland

The adrenal glands are situated on the cranial surface of the kidneys. There are two parts to this endocrine gland, an outer cortex and an inner medulla.

The adrenal cortex produces several hormones. These include:

1. Aldosterone that regulates the concentration of sodium and potassium in the blood by controlling the amounts that are secreted or reabsorbed in the kidney tubules.
2. Cortisone and hydrocortisone (cortisol) that have complex effects on glucose, protein and fat metabolism. In general they increase metabolism. They are also often administered to animals to counteract allergies and for treating arthritic and rheumatic conditions. However, prolonged use should be avoided if possible as they can increase weight and reduce the ability to heal.
3. Male and female sex hormones similar to those secreted by the ovaries and testes.

The hormones secreted by the adrenal cortex also play a part in "general adaptation syndrome" which occurs in situations of prolonged stress.

The adrenal medulla secretes adrenalin (also called epinephrine). Adrenalin is responsible for the so-called flight fight, fright response that prepares the animal for emergencies. Faced with a perilous situation the animal needs to either fight or make a rapid escape. To do either requires instant energy, particularly in the skeletal muscles. Adrenaline increases the amount of blood reaching them by causing their blood vessels to dilate and the heart to beat faster. An increased rate of breathing increases the amount of oxygen in the blood and glucose is released from the liver to provide the fuel for energy production. Sweating increases to keep the muscles cool and the pupils of the eye dilate so the animal has a wide field of view. Functions like digestion and urine production that are not critical to immediate survival slow down as blood vessels to these parts constrict.

Note that the effects of adrenalin are similar to those of the sympathetic nervous system.

The Pancreas

In most animals the pancreas is an oblong, pinkish organ that lies in the first bend of the small intestine. In rodents and rabbits, however, it is spread thinly through the mesentery and is sometimes difficult to see.

Most of the pancreas acts as an exocrine gland producing digestive enzymes that are secreted into the small intestine. The endocrine part of the organ consists of small clusters of cells (called Islets of Langerhans) that secrete the hormone insulin. This hormone regulates the amount of glucose in the blood by increasing the rate at which glucose is converted to glycogen in the liver and the movement of glucose from the blood into cells.

In diabetes mellitus the pancreas produces insufficient insulin and glucose levels in the blood can increase to a dangerous level. A major symptom of this condition is glucose in the urine.

The Ovaries

The ovaries, located in the lower abdomen, produce two important sex hormones.

1. The follicle cells, under the influence of FSH (see the pituitary gland above), produce oestrogen, which stimulates the development of female sexual characteristics - the mammary glands, generally smaller build of female animals etc. It also stimulates the thickening of the lining of the uterus in preparation for pregnancy (13).
2. Progesterone is produced by the corpus luteum, the endocrine gland that develops in the empty follicle following ovulation (13). It promotes the further preparation of the uterine lining for pregnancy and prevents the uterus contracting until the baby is born.

The Testes

Cells around the sperm producing ducts of the testis produce the hormone testosterone. This stimulates the development of the male reproductive system and the male sexual characteristics - generally larger body of male animals, mane in lions, tusks in boars, etc.

8

Animal Breeding

Animal breeding is a branch of animal science that addresses the evaluation (using best linear unbiased prediction and other methods) of the genetic value (estimated breeding value, EBV) of domestic livestock. Selecting animals for breeding with superior EBV in growth rate, egg, meat, milk, or wool production, or have other desirable traits has revolutionized agricultural livestock production throughout the world. The scientific theory of animal breeding incorporates population genetics, quantitative genetics, statistics, and recently molecular genomics and is based on the pioneering work of Sewall Wright, Jay Lush, and Charles Henderson.

BREEDING STOCK

Breeding stock is a group of animals used for purpose of planned breeding. When individuals are looking to breed animals, they look for certain valuable traits in purebred animals, or may intend to use some type of crossbreeding to produce a new type of stock with different, and presumably superior abilities in a given area of endeavor. For example, when breeding swine the "breeding stock should be sound, fast growing, muscular, lean, and reproductively efficient." The "subjective selection of breeding stock" in horses has led to many horse breeds with particular performance traits.

Purebred breeding

Mating animals of the same breed for maintaining such breed is referred to as purebred breeding. Opposite to the practice of mating animals of different breeds, purebred breeding aims to establish and maintain stable traits, that animals will pass to the next generation. By "breeding the best to the best," employing a certain degree of inbreeding, considerable culling, and selection for "superior" qualities, one could develop a bloodline or "breed" superior in certain respects to the original base stock. Such animals can be recorded with a breed registry, the organisation that maintains pedigrees and/or stud books. The observable phenomenon of hybrid vigor stands in contrast to the notion of breed purity.

Backyard breeding

In the United States, a backyard breeder is someone who breeds animals, often without registration and with a focus on profit. In some cases the animals are inbred narrowly for looks with little regard to health. The term is considered derogatory. If a backyard dog breeder has a significant number of breeding animals, they become associated with puppy mills.

Artificial Insemination

Artificial insemination (AI) is the deliberate introduction of semen into a female's vagina or oviduct for the purpose of achieving a pregnancy through fertilisation by means other than copulation. It is the medical alternative to sexual intercourse, or natural insemination.

Artificial insemination is a fertility treatment for humans, and is a common practice in the breeding of dairy cattle and pigs. Artificial insemination may employ assisted reproductive technology, donated sperm, and/or animal husbandry techniques.

Artificial insemination is the technique in which semen with living sperms is collected from the male and introduced into female reproductive tract at proper time with the help of instruments. This has been found to result in a normal offspring. In this process, the semen is inseminated into the female by placing a portion of it either in a collected or diluted form into the cervix or uterus by mechanical methods at the proper time and under most hygienic conditions. The first scientific research in artificial insemination of domestic animals was performed on dogs in 1780 by the Italian scientist, Lazanno Spalbanzani. His experiments proved that the fertilizing power reside in the spermatozoa and not in the liquid portion of semen. Few further studies under research station conditions helped this technique to be used commercially allover the world including India.

Artificial insemination is not merely a novel method of bringing about impregnation in females. Instead, it is a powerful tool mostly employed for livestock improvement. In artificial insemination the germplasm of the bulls of superior quality can be effectively utilized with the least regard for their location in far away places. By adoption of artificial insemination, there would be considerable reduction in both genital and non-genital diseases in the farm stock.

SYMPTOMS OF HEAT

The various symptoms of heat are

- The animal will be excited condition. The animal will be in restlessness and nervousness.
- The animal will be bellow frequency.
- The animal will reduce the intake of feed.
- Peculiar movement of limbo sacral region will b observed.
- The animals which are in heat will lick other animals and smelling other animals.
- The animals will try to mount other animals
- The animals will standstill when other animal try to mount.. This period is known as standing heat. This extends 14-16 hours.
- Frequent maturation (urination) will be observed.
- Clear mucous discharge will be seen from the vulva, sometimes it will be string like the mucous will be seen stick to the near the pasts of valva.
- Swelling of the valva will be seen.
- 11 Congestion and hyperemia of membrane.
- The tail will be in raised position.
- Milk production will be slightly decreased.
- On Palpation uterus will be turgid and the cervix will be opened.

A.I. ADVANTAGES - DISADVANTAGES

Advantages and disadvantages:

Artificial insemination (A.I.) is deposition of semen into the female genital tract by means of instruments.

Advantages of artificial insemination

There are several advantages by artificial insemination over natural mating or servicing.

- There is no need of maintenance of breeding bull for a herd; hence the cost of maintenance of breeding bull is saved.
- It prevents the spread of certain diseases and sterility due to genital diseases.
- Eg: contagious abortion, vibriosis.
- By regular examination of semen after collection and frequent checking on fertility make early detection of interior males and better breeding efficiency is ensured.
- The progeny testing can be done at an early age.
- The semen of a desired size can be used even after the death of that particular sire.
- The semen collected can be taken to the urban areas or rural areas for insemination.
- 7 It makes possible the mating of animals with great differences in size without injury to either of the animal.
- It is helpful to inseminate the animals that are refuse to stands or accept the male at the time of oestrum.
- It helps in maintaining the accurate breeding and cawing records.
- It increases the rate of conception.
- It helps in better record keeping.
- Old, heavy and injured sires can be used.

Disadvantages of A.I

- Requires well-trained operations and special equipment.
- Requires more time than natural services.
- Necessitates the knowledge of the structure and function of reproduction on the part of operator.
- Improper cleaning of instruments and in sanitary conditions may lead to lower fertility.
- If the bull is not properly tested, the spreading of genital diseases will be increased.
- Market for bulls will be reduced, while that for superior bull is increased.

Semen Collection Methods And Evaluation

Various methods of collection of semen have been devised from time to time. The older unsatisfactory methods have gradually replaced by the new modern techniques.

There are three common methods.

1. Use of artificial vagina
2. By Electro-stimulation method.
3. By massaging the ampulae of the duct us differences through rectal wall.
4. The ideal method of semen collection is use of artificial vagina which is safe for sire and the collector also.

Artificial Vagina Method:

The artificial vagina has the following parts:

- A heavy hard rubber 2″ lose, open at both ends with a nostle for air and water in and outlet.
- Inner sleeve of rubber or rubber liner.
- The semen receiving cone or rubber cone.
- Semen collection tube made of glass or plastic graduate in cc and its fraction correct to 0.1 CC
- Insulating bag Before using for semen collection all the parts are washed thoroughly and sterilized properly, and assembled as artificial vagina, the rubber liner is inserted into the hose; inverting both ends back by folding back from either side opening, and fastening with rubber bands. Now the space between the hard rubber hose and inner rubber liner forms a water tight compartment. The nostle at one end of the hose can be fixed .

PARTS OF ARTIFICIAL VAGINA

Turning through the threaded nut up or down. The water jacket of the Artificial -vagina is- filled with hot water at a temperature of 45 C (113 F) by opening the nostle. The graduated semen collection tube is fixed to the narrow end of the artificial vagina hose, and fastened by a rubber band. The inner side of the rubber liner on the anterior side of the artificial vagina is lubricated with sterile jelly to a length of 3 to 4 inches. Air is blown through the nostle into the water jacket, to create pressure in if, and the same is exerted the rubber linear, to simulate natural vagina.

The temperature of the artificial vagina is to be checked, at each collection, and it should simulate natural vagina at mounting time. If the artificial vagina is to mount later. If it is too cold ejaculate may not be there after a thrust, or even if ejaculate is there; it may be contaminated with urine, and becomes unfit for use.

Semen Collection Method. (A.v.)

The cow or dummy is secured in service create. The artificial vagina assembled is held at 45 angle from the direction of penis, and the thrust is that angle. The artificial vagina is held with the left hand by a right handed person; and when the bull mounts the cow, the sheath of the bull will be graphed by the operator, directing the gland penis into the artificial vagina, and then the bull gives a thrust to ejaculate. The operator should evince care so as not to touch the exposed past of the penis. After the bull dismounts, the artificial vagina is taken off from penis and the air vent is opened to release the pressure from the jacket. The water from the jacket is also drained by opening the nostle. This allows the ejaculate to flow from the cone to the semen collection tube. The semen collection tube is detached from the cone, plugged with cotton wool, and taken to the laboratory for examination. The rubber cone and the semen collection tube can be protected from external contamination or heat or higher, by covering with an insulation bag with zip.

Semen Storage

The discovery that bull semen could be successfully frozen and stored for indefinite periods has revolutionized AI in cattle. In 1949, British scientists discovered that addition of glycerol to the semen extender improved resistance of sperm to freezing. Glycerol acts to remove water from the sperm cell prior to freezing and prevents the formation of cellular ice crystals which would damage the sperm. There are two methods of freezing and storing semen: dry ice and alcohol (-100 degrees F) and liquid nitrogen (-320 degrees F). Liquid nitrogen is preferred because there is no evidence of fertility deterioration with age. Fertility gradually declines in semen stored in dry ice-alcohol.

Frozen semen can be stored indefinitely if proper temperature is maintained. A recent report told of a calf born from frozen semen stored for 16 years. Fresh, liquid semen can be successfully stored for 1 to 4 days at 40 degrees F. Semen is usually stored in glass ampoules. Other methods appear promising, particularly the French-straw. Several AI organizations have gone to this method exclusively. Artificial coloring is frequently added to semen extenders in order to distinguish one breed from another. Complete identification of the bull is required on each individual semen container.

INSEMINATION METHODS

There-are different methods insemination in different species of animals i.e. speculum method, vaginal method and recto vaginal method.

Recto Vaginal Method

In cattle the safe and best method of insemination is "Recto vaginal method of insemination". Cow which is in heat is well controlled placing it in a Travis. The inseminator will get ready by wearing a plastic apron, gumboots and gloves. The semen straw after thawing (keeping the semen straw in warm water for a minute to convert the freezed semen into liquid and the sperms become motile) is loaded in a sterilized A.I. gum and is covered with a plastic sheath. The inseminator will insert the gloved left hand into the rectum after applying the soft soap or other lubricant on the glove and back racked the animal, and the hand is further inserted and will catch hold the cervix through rectal wall. The A.I gum loaded with semen straw is passed.

Recto-vaginal method of insemination

Through the vulva to 'vagina and cervix and observed with the hand in rectum that the A. I gum reaches the cervix, then the semen is deposited by injecting the gun, and after depositing the semen the gun is removed, the empty straw and sheath are disordered.

Spectrum Method

In this method spectrum is placed in the vagina of the cow, which provides passage outside to the site of insemination, then inseminating tube is passed through the speculum and semen is deposited at the cervix insemination method.

Vaginal Method

Hand is passed through the vagina and the inseminating tube is guided by hand to the site of insemination and semen is deposited. Here there is a risk of contamination and injury of female genitalia.

FROZEN SEMEN AND STORAGE

Freezing of semen for successful preservation of spermatozoa, for long periods, is of great importance in livestock breeding and farm management. It has made it possible" to make available the use of outstanding proven sizes for larger number of cows, covering larger area, frozen semen shipment has become possible to different continents in the globe to any place connected with any service. Now a day if farmer wants to use of an outstanding size for inheritance of high milk yield, he can go in for frozen semen service provided his area is, covered by Artificial insemination, with supply of frozen semen.

At present frozen semen is used in most of the states in India. The technique of semen preservation in straws was developed in France . Freezing of semen is done with a special diluents, which has the following composition.

Sodium citrate dihydrate (angular) 2.4 y. 2.0 gm 8.0 ml 25.0%byvolume 50,000 units per 100 ml of semen Fructose Glycerol Egg Yolk Penicillin dilulent. Dihydro-streptomycin 50.0 mg per .100m1 of semen dilulent. Distilled water double glass distilled 100.Om1. The addition of glycerol to the dilulent makes the cells more resistant to the rigours of freezing and icy crystals, which form are smaller and smoother thus creating less damage to the spermatozoa. The addition of fructose to the diluent luprores sperm resistance to glycerol; and also provides nutrition.

Frozen semen is packed in single dose glass vials or plastic straws at +5 C. The final level of glycerol should be 7.0 to 7.6% during the freezing process. The antibiotics are added to inhibit bacteria and to kill pathetic organisms. The semen to be diluted in such a way that one ml. of extended semen will contain 20 million motile spermatozoa. The semen must be cooled carefully for spermatozoa to remain with life. The final temperature is lowered to -79 C or still lower. Quick freezing is done for a period of 3 to 5 minutes to -75 C with the help of atmosphere created by liquid nitrogen. In the slow freezing technique cooling is done at the rate of 1 C per minute from +5 C to -15 C. From -15 C to -31 C at the rate of 2 C per minute. From -31 C to 75 C at the rate of 4 to 5 C per minute. Thus taking 40 minutes in total, further cooling to -96 C can be done quickly as it is not critical after freezing. Before freezing the diluted semen in equilibreated for 3 to 5 hours or for the best 16 to 20, hours period in refrigerator at 5 C. Frozen semen facilitates the percent use of the semen diluted and frozen, and thus the delivery price is reduced, and it can be supplied with the gaps of months to the A.I technicians as against the supply of fluid semen every days or alternate days. Liquid nitrogen plays a vital role for storing the frozen semen straws, at a temperature of -196 C for longer periods.

Artificial insemination (AI) of cattle

Artificial insemination (AI) is the process of collecting sperm cells from a male animal and manually depositing them into the reproductive tract of a female. One can cite a number of potential benefits from the use of artificial insemination.

Increased efficiency of bull usage:

During natural breeding, a male will deposit much more semen than is theoretically needed to produce a pregnancy. In addition, natural breeding is physically stressful. Both of these factors limit the number of natural mating a male can make. However, collected semen can be diluted and extended to create hundreds of doses from a single ejaculate. Also, semen can be easily transported; allowing multiple females in different geographical locations to be inseminated simultaneously, and semen can be stored for long periods of time, meaning that males can produce offspring long after their natural reproductive lives end.

Increased potential for genetic selection:

Because artificial insemination allows males to produce more offspring, fewer males are needed. Therefore, one can choose only the few best males for use as parents, increasing the selection intensity. Furthermore, because males can have more offspring, their offspring can be used in a progeny test program to more accurately evaluate the genetic value of the male. Finally, individual farmers can use artificial insemination to increase the genetic pool with which his or her animals can be mated, potentially decreasing effects of inbreeding.

DECREASED COSTS

Male animals often grow to be larger than females and can consume relatively larger amounts of feed. Also, male animals are often more strong, powerful, and potentially ill-mannered and thus require special housing and handling equipment.

Increased safety for animals and farmers:

As mentioned, male animals can become large and aggressive. These factors mean that maintaining a bull on a farm may be dangerous. Also, because of the relatively larger size of adult males than females, natural mating is more likely to result accidents and injury to either the cow or the bull than is artificial insemination.

Timing of Insemination for Maximum Conception

A frequent question concerning AI is: What time during estrus should cows be bred for greatest chance of conception? Since estrus may last from 10 to 25 hours there is considerable latitude in possible time of insemination. Much research work has been conducted on this subject.

Controlled investigations were conducted by Trim Berger and Davis at Nebraska in 1943. These and other studies show that conception rate is lower when cows are bred prior to mid estrus or later than 6 hours after cessation of estrus (standing heat in this case). Maximal conception is obtained when cows are inseminated between mid estrus and the end of standing estrus, with good results up to 6 hours after estrus.

Success in insemination timing is dependent upon a good heat detection program. In large herds, this means assigning individual responsibility for heat detection and a continued education program for labor. A successful heat detection program and subsequent proper timing of insemination will pay dividends in increasing reproductive efficiency.

9

Animal Husbandry in India

All living things are divided into seven categories using Latin terms that can be understood worldwide. They progressively get smaller and more numerous, and each contains less living things. Classification of living things is used to help identify different animals and to group them together with their relatives.

The first and largest category is the Kingdom. To date there are five kingdoms: Animalia, which is made up of animals; Plantae, which is made up of plants; Protista, which is made up of protists (single-celled creatures invisible to the human eye); Fungi, which is made up of mushrooms, mold, yeast, lichen, etc; and Monera, which is made up of the three types of bacteria.

The next category is the Phylum. There are several phyla within each kingdom. The phyla start to break the animals (or plants, fungi, etc) into smaller and more recognizable groups. The best known phylum is Chordata, which contains all animals with backbones (fish, birds, mammals, reptiles, amphibians). There is also Arthropoda (insects, spiders, crustaceans); Mollusca (snails, squid, clam); Annelida (segmented worms); Echinodermata (starfish, sea urchins) and many, many more.

The next category that makes up the phyla is the Class. The class breaks up animals into even more familiar groups. For example, the phylum Chordata is broken down into several classes, including Aves (birds), Reptilia (reptiles), Amphibia (amphibians), Mammalia (mammals) and several others.

The next category is the Order. Each class is made up of one or more orders. Mammalia can be broken down into Rodentia (mice, rats), Primates (Old- and New-World monkeys), Chiroptera (bats), Insectivora (shrews, moles), Carnivora (dogs, cats, weasels), Perissodactyla (horses, zebras), Artiodactyla (cows), Proboscidea (elephants) and many more.

Orders can then be broken down into Families. The order Carnivora can be broken down into Canidae (dogs), Felidae (cats), Ursidae (bears), Hyaenidae (hyaenas, aardwolves), Mustelidae (weasels, wolverines), and many more.

The next category is the Genus. The family Felidae, for example, can be broken down into Acinonyx (cheetah), Panthera (lion, tiger), Neofelis (clouded leopard) and Felis (domestic cats).

Finally, the genus is broken down into the Species. The genus Panthera can be broken down to include Panthera leo (lion) and Panthera tigris (tiger). Note that the genus is placed in front of the species. To complicate things, each category can, in some cases, be broken down into sub-categories. A sub-category is made when there are some animals that are basically the same but live in different areas and behave differently from each other or have some general differences in the anatomy. A prime example is phylum Chordata, class Reptilia, order Squamata. The order

Squamata consists of both lizards and snakes because they are basically the same. However, snakes have some physical differences, mainly the lack of legs (although some lizards also lack legs). This warrants the creation of two sub-orders: Serpentes (snakes) and Sauria (lizards). These are then broken down into their own families.

Another example is phylum Chordata, class Mammalia, order Proboscidea, family Elephantidae, genus Elephas and species maximus. Elephas maximus (Asian elephant) is broken down into four sub-species depending on where the sub-species lives and the physical description of it. One example is the Ceylon elephant (Elephas maximus maximus). Note that the sub-species follows the species and that they always receive lower-case letters.

It should be noted that most animals are given common names recognized outside of the scientific community and that these names may be different in each country. For example, in Canada a certain type of spider is known as the daddy long-legs and is known as the harvester spider in the US. Likewise a type of fly in the US is called the daddy long-legs and is known as the crane fly in Canada. However, the Latin term remains the same.

It should lastly be noted that the Latin names of animals are always changing. The Aleutian Canada goose has gone through ten name changes, starting off as the non-sub-species Anser leucopreius and ending with the sub-species Branta canadensis leucopareia. The name will most likely change again in the future if it is decided that the Aleutian Canada goose is an entirely separate species from the Canada goose or if there are no subspecies at all and just the Canada goose.

Definition

The exact definition of taxonomy varies from source to source, but the core of the discipline remains: the conception, naming, and classification of organism groups. The exact relationship of systematics and classification to taxonomy also varies because the usage of the terms in biology originated independently. As points of reference, recent definitions of taxonomy are presented below:

1. Theory and practice of grouping individuals into species, arranging species into larger groups, and giving those groups names, thus producing a classification;
2. A field of science (and major component of systematics) that encompasses description, identification, nomenclature, and classification; see
3. The science of classification, in biology the arrangement of organisms into a classification.
4. "The science of classification as applied to living organisms, including study of means of formation of species, etc."
5. "The analysis of an organism's characteristics for the purpose of classification"

The varied definitions either place taxonomy as a sub-area of systematics (definition 2), or appear to consider the two terms synonymous. There is some disagreement as to whether biological nomenclature is considered a part of taxonomy (definitions 1 and 2), or a part of systematics outside taxonomy. For example, the last definition is paired with the following definition of systematics that places nomenclature outside taxonomy:

- Systematics: "The study of the identification, taxonomy and nomenclature of organisms, including the classification of living things with regard to their natural relationships and the study of variation and the evolution of taxa".

  History of taxonomy

  Early taxonomists

Taxonomy has been called "the world's oldest profession", and naming and classifying our surroundings has likely been taking place as long as mankind has been able to communicate. It would always have been important to know the names of poisonous and edible plants and animals in order to communicate this information to other members of the family or group.

In the East, one of the earliest recorded pharmacopoeias was written by Shen Nung, Emperor of China (c. 3000 BC). He wanted to spread information related to agriculture and medicine, and is said to have tasted hundreds of plants with the goal of learning their medicinal value. Records after this are difficult to interpret for some time, but medicinal plant illustrations show up in Egyptian wall paintings from c. 1500 BC. The paintings clearly show that these societies valued and communicated the uses of different species, and therefore had a basic taxonomy in place.

ARISTOTLE TO PLINY THE ELDER

Historical records show that informally classifying organisms took place at least back to the days of Aristotle (Greece, 384-322 BC), who was the first to begin to classify all living things. Some of the terms he gave to animals, such as "invertebrates" and "vertebrates" are still commonly used today. His student Theophrastus (Greece, 370-285 BC) carried on this tradition, and wrote a classification of 480 plants called Historia Plantarum. Again, several plant groups currently still recognized can be traced back to Theophrastus, such as Cornus, Crocus, and Narcissus. The next major turn-of-the-millennia era taxonomist came in the form of Pliny the Elder (Rome, 23-79 AD). His elaborate 160-volume work Naturalis Historia described many plants, and even gave many of them Latin binomial names.

Pre-Linnaean taxonomists

It was not until c. 1500 years later that taxonomic works became ambitious enough to replace the ancient texts. This is often credited to the development of sophisticated optic lenses, which allowed for the morphology of organisms to be studied in much greater detail. One of the earliest authors to take advantage of this leap in technology was Andrea Cesalpino (Italy, 1519-1603), who is often referred to as "the first taxonomist". His magnum opus De Plantis came out in 1583, and described over 1500 plant species. Two large plant families that he first recognized are still in use today: the Asteraceae and Brassicaceae. Then in the seventeenth century John Ray (England, 1627-1705) wrote many important taxonomic works. Arguably his greatest accomplishment was Methodus Plantarum Nova (1682), where he published over 18,000 plant species. At the time his classifications were perhaps the most complex yet produced by any taxonomist, as he based his taxa on many combined characters. The next major taxonomic works were produced by Joseph Pitton de Tournefort (France, 1656-1708). His work from 1700, Institutiones Rei Herbariae, included over 9000 species in 698 genera, and directly influenced Linnaeus as it was the text he used as a young student.

The Linnaean era

The Swedish botanist Carl Linnaeus (1707-1778) ushered in a new era of taxonomy. With his major works Systema Naturae 1st Edition in 1735, Species Plantarum in 1753, and Systema Naturae 10th Edition, he revolutionized modern taxonomy. His works implemented a standardized binomial naming system for animal and plant species, which proved to be an elegant solution to a chaotic and disorganized taxonomic literature. As a result the Linnaean system was born, and is still used in essentially the same way today as it was in the eighteenth century. Currently, plant and animal taxonomists regard Linnaeus' work as the "starting point" for valid names (at 1753 and 1758 respectively). Names published before these dates are referred to as "pre-Linnaean", and not considered valid (with the exception of spiders published in Svenska Spindlar). Even taxonomic names published by Linnaeus himself before these dates are considered pre-Linnaean.

The word taxonomy was introduced in 1813 by Candolle, in his Théorie élémentaire de la botanique.

PHYLOGENETICS AND CLADISTICS

Today, traditional rank-based biological classifications persist in a structure largely unchanged since the 1700s; however, how the relationships of these taxa are investigated has changed drastically in recent decades. It is now common for biologists to devise a classification based on the results of phylogenetic analysis using DNA sequence data, and taxa are typically required to be clades. Although phylogenetics itself is fundamental to modern-day systematics, its use for the description of new taxa, and for their placement within a classification scheme, is not required.

Application

Biological taxonomy is a sub-discipline of biology, and is generally practiced by biologists known as "taxonomists", though enthusiastic naturalists are also frequently involved in the publication of new taxa. The work carried out by taxonomists is crucial for the understanding of biology in general. Two fields of applied biology in which taxonomic work is of fundamental importance are the study of biodiversity and conservation. Without a working classification of the organisms in any given area, estimating the amount of diversity present is unrealistic, making informed conservation decisions impossible. As conservation becomes ever more politically important, taxonomic work impacts not only the scientific community, but society as a whole.

Classifying organisms

Biological classification is a critical step in the taxonomic process, as it informs the user as to what the relatives of the taxon are hypothesized to be. Although the discipline of taxonomy itself does not deal with the investigations of how taxa are related to one another, it does serve to communicate these results to the user. To do this, it uses taxonomic ranks, including, among others (in order from most inclusive to least inclusive): Domain, Kingdom, Phylum, Class, Order, Family, Genus, and Species.

Taxonomic descriptions

The 'definition' of a taxon is encapsulated by its description. There are no set rules governing the definition of taxa, but the naming and publication of new taxa is governed by sets of rules. In zoology, the nomenclature for the more commonly used ranks (superfamily to subspecies), is regulated by the International Code of Zoological Nomenclature (ICZN Code). In the fields of botany, phycology, and mycology, the naming of taxa is governed by the International Code of Nomenclature for algae, fungi, and plants (ICN).

The initial description of a taxon involves five main requirements:

1. The taxon must be given a name based on the 26 letters in the Latin alphabet (a binomial for new species, or uninomial for other ranks).
2. The name must be unique (i.e. not a homonym).
3. The description must be based on at least one name-bearing type specimen.
4. It should include reference to enough attributes to differentiate the taxon from other taxa (ICZN Code, Article 13.1.1; ICN, Article 38.2). Both codes deliberately separate defining the content of a taxon (its circumscription) from defining its name.
5. These first four requirements must be published in a work that is obtainable in numerous identical copies, as a permanent scientific record.

However, often much more information is included, like the geographic range of the taxon, ecological notes, chemistry, behavior, etc. How researchers arrive at their taxa varies; depending on the available data, and resources, methods vary from simple quantitative or qualitative comparisons of striking features, to elaborate computer analyses of large amounts of DNA sequence data.

PHENETICS

In phenetics, also known as taximetrics, organisms are classified based on overall similarity, regardless of their phylogeny or evolutionary relationships. It results in a measure of evolutionary "distance" between taxa. Phenetic methods have become relatively rare in modern times, largely superseded by cladistic analyses, as phenetic methods do not distinguish plesiomorphic from apomorphic traits. However, certain phenetic methods, such as neighbor joining, have found their way into cladistics, as a reasonable approximation of phylogeny when more advanced methods (such as Bayesian inference) are too computationally expensive.

Databases

Modern taxonomy uses database technologies to search and catalog classifications and their documentation. While there is no commonly used database, there are comprehensive databases such as the Catalogue of Life, which attempts to list every documented species. The catalogue listed 1.4 million species for all kingdoms as of May 2012, claiming coverage of more than 74% of the estimated 1.9 million species known to modern science.

Domestication

Domestication (from Latin domesticus) is the process whereby a population of living organisms is changed at the genetic level, through generations of selective breeding, to accentuate traits that ultimately benefit humans. A usual by-product of domestication is the creation of a dependency in the domesticated organisms, so that they lose their ability to live in the wild. It differs from taming in that a change in the phenotypical expression and genotype of the animal occurs, whereas taming is simply the process by which animals become accustomed to human presence. In the Convention on Biological Diversity, a domesticated species is defined as a "species in which the evolutionary process has been influenced by humans to meet their needs." Therefore, a defining characteristic of domestication is artificial selection by humans. Humans have brought these populations under their control and care for a wide range of reasons: to produce food or valuable commodities (such as wool, cotton, or silk) and for types of work (such as transportation, protection, warfare), scientific research, or simply to enjoy as companions or ornaments.

Plants domesticated primarily for aesthetic enjoyment in and around the home are usually called house plants or ornamentals, while those domesticated for large-scale food production are generally called crops. A distinction can be made between those domesticated plants that have been deliberately altered or selected for special desirable characteristics (see cultigen) and those plants that are used for human benefit, but are essentially no different from the wild populations of the species. Animals domesticated for home companionship are usually called pets while those domesticated for food or work are called livestock or farm animals.

Background

Charles Darwin was the first to describe how domestication, selection and evolution are interlinked, and based on natural heritable variation among individual plants and animals. Today we know that such natural variation is caused by mutations in genes coding for these traits, and by new combinations of already existing genetic variation, based on earlier mutations. Darwin described how the process of domestication can involve both unconscious and methodical elements. Routine human interactions with animals and plants create selection pressures that cause adaptation to human presence, use or cultivation. Deliberate selective breeding has also been used to create desired changes, often after initial domestication. These two forces, unconscious natural selection and methodical selective breeding, may have both played roles in the processes of domestication throughout history. Both have been described from man's perspective as processes of artificial selection.

The domestication of wheat provides an example. Wild wheat falls to the ground to reseed itself when ripe, but domesticated wheat stays on the stem for easier harvesting. There is evidence that this change was possible because of a random mutation that happened in the wild populations at the beginning of wheat's cultivation. Wheat with this mutation was harvested more frequently and became the seed for the next crop. Therefore, without realizing, early farmers selected for this mutation, which may otherwise have died out. The result is domesticated wheat, which relies on farmers for its own reproduction and dissemination.

The domestication of dogs provides another example. It is speculated that tamer than average wolves, less wary of humans, selected themselves as dogs over many generations. These wolves were able to thrive by following humans to scavenge for food near camp fires and garbage dumps, which gave them an advantage over more shy individuals. Eventually a symbiotic relationship developed between people and these proto-dogs. The dogs fed on human food scraps, and humans found that dogs could warn them of approaching dangers, help with hunting, act as pets, provide warmth, or supplement their food supply. As this relationship progressed, humans eventually began to keep these self-tamed wolves and breed from them the types of dogs that we have today.

In recent times, selective breeding may best explain how continuing processes of domestication often work. Some of the best-known evidence of the power of selective breeding comes from an experiment by Russian scientist, Dmitri K. Belyaev, in the 1950s. His team spent many years breeding the Silver Fox (Vulpes vulpes) and selecting only those individuals that showed the least fear of humans. Eventually, Belyaev's team selected only those that showed the most positive response to humans. He ended up with a population of grey-coloured foxes whose behavior and appearance was significantly changed. They no longer showed any fear of humans and often wagged their tails and licked their human caretakers to show affection. These foxes had floppy ears, smaller skulls, rolled tails and other traits commonly found in dogs.

Despite the success of this experiment, it appears that selective breeding cannot always achieve domestication. Attempts to domesticate many kinds of wild animals have been unsuccessful. The zebra is one example. Despite the fact that four species of zebra can interbreed with and are part of the same genus as the horse and the donkey, attempts at domestication have failed. Factors such as temperament, social structure and ability to breed in captivity play a role in determining whether a species can be successfully domesticated. In human history to date, only a few species of large animal have been domesticated. In approximate order of their earliest domestication these are: dog, sheep, goat, pig, ox, yak, reindeer, water buffalo, horse, donkey, llama, alpaca, Bactrian camel and Arabian camel.

Animals

1. According to evolutionary biologist Jared Diamond, animal species must meet six criteria in order to be considered for domestication:Flexible diet - Creatures that are willing to consume a wide variety of food sources and can live off less cumulative food from the food pyramid (such as corn or wheat), particularly food that is not utilized by humans (such as grass and forage) are less expensive to keep in captivity. Carnivores by definition feed primarily or only on animal tissue, which requires the expenditure of many animals, though they may exploit sources of meat not utilized by humans, such as scraps and vermin.
2. Reasonably fast growth rate - Fast maturity rate compared to the human life span allows breeding intervention and makes the animal useful within an acceptable duration of caretaking. Large animals such as elephants require many years before they reach a useful size.
3. Ability to be bred in captivity - Creatures that are reluctant to breed when kept in captivity do not produce useful offspring, and instead are limited to capture in their wild state. Creatures such as the panda, antelope and giant forest hog are territorial when breeding and cannot be maintained in crowded enclosures in captivity.

4. Pleasant disposition - Large creatures that are aggressive toward humans are dangerous to keep in captivity. The African buffalo has an unpredictable nature and is highly dangerous to humans; similarly, although the American bison is raised in enclosed ranges in the Western United States, it is much too dangerous to be regarded as truly domesticated. Although similar to the domesticated pig in many ways, Africa's warthog and bushpig are also dangerous in captivity.
5. Temperament which makes it unlikely to panic - A creature with a nervous disposition is difficult to keep in captivity as it may attempt to flee whenever startled. The gazelle is very flighty and it has a powerful leap that allows it to escape an enclosed pen. Some animals, such as the domestic sheep, still have a strong tendency to panic when their flight zone is encroached upon. However, most sheep also show a flocking instinct, whereby they stay close together when pressed. Livestock with such an instinct may be herded by people and dogs.
6. Modifiable social hierarchy - Social creatures whose herds occupy overlapping ranges and recognize a hierarchy of dominance can be raised to recognize a human as the pack leader:
   1. tapirs and rhinoceroses are solitary and do not tolerate being penned with each other
   2. antelope and deer except for reindeer are territorial when breeding and live in herds only for the rest of the year
   3. bighorn sheep and peccaries have nonhierarchical herd structures and do not follow any definite leader: instead males fight continuously with each other for mating opportunities
   4. musk ox herds (although having a defined leader) maintain mutually exclusive territories and two herds will fight if kept together.

However, this list is of limited use because it fails to take into account the profound changes that domestication has on a species. While it is true that some animals, including parrots, whales, and most members of the Carnivora, retain their wild instincts even if born in captivity, some factors must be taken into consideration.

In particular, number (5) may not be a prerequisite for domestication, but rather a natural consequence of a species' having been domesticated. In other words, wild animals are naturally timid and flighty because they are constantly faced by predators; domestic animals do not need such a nervous disposition, as they are protected by their human owners. The same holds true for number (4) - aggressive temperament is an adaptation to the danger from predators. A Cape buffalo can kill even an attacking lion, but most modern large domestic animals were descendants of aggressive ancestors. The wild boar, ancestor of the domestic pig, is certainly renowned for its ferocity; other examples include the aurochs (ancestor of modern cattle), horse, Bactrian camels and yaks, all of which are no less dangerous than their undomesticated wild relatives such as zebras and buffalos. Others have argued that the difference lies in the ease with which breeding can improve the disposition of wild animals, a view supported by the failure to domesticate the kiang and onager. On the other hand for thousands of years humans have managed to tame dangerous species like the elephants, bears and cheetahs whose failed domestications had little to do with their aggressiveness.

Number (6), while it does apply to most domesticated species, also has exceptions, most notably in the domestic cat and ferret, which are both descended from strictly solitary wild ancestors but which tolerate and even seek out social interaction in their domestic forms. Feral domestic cats, for example, naturally form colonies around concentrated food sources and will even share prey and rear kittens communally, while wildcats remain solitary even in the presence of such food sources. Zoologist Marston Bates devoted a chapter on domestication in his 1960 book The Forest and the Sea, in which he talks a great deal about how domestication alters a species: Dispersal mechanisms tend to disappear for the reason stated above, and also because people provide transportation for them. Chickens have practically lost their ability to fly. Similarly, domestic animals cease to have a definite mating season, and so the need to be territorial when mating loses its value; and if some of the males in a herd are castrated, the problem is reduced even further. What he says suggests that the process of domestication can itself make a creature domesticable. Besides, the first steps towards agriculture may have involved hunters keeping young animals, who are always more impressionable than the adults, after killing their mothers.

Another strong factor deciding whether a species will be considered for domestication is quite simply the availability of more suitable (or even better already domesticated) alternatives. For example a community that had been introduced to domestication by neighboring peoples will generally find it much more practical, economical and time saving to import already domesticated species than experiment with wild animals (even if they are of the same species). Generally speaking, the species of animals originally domesticated by early humans in the interconnected landmasses of Eurasia and Africa were far superior, both in working capacity and in food production, than the species found in the other continents, namely the Americas and Oceania.

HISTORY OF THE DOMESTICATION OF ANIMALS

Hunting species: 20,000-10,000 years ago

During the most recent ice age, from about 20,000 years ago (see Ice Ages), large mammals such as bison roam on the sub-arctic tundra of Europe and Asia. They are preyed upon by two groups of hunters, both much smaller and weaker than themselves - but both with a sufficiently developed social system to enable them to hunt and kill in packs.

These hunters are humans and wolves. The typical pack of wolves and of humans is surprisingly similar. It is family-based, led by a dominant male whose female partner is likely to have an authority second only to his. Members of the group are friendly to each other but deeply suspicious of outsiders. All members (not just the parents) are protective of the newly born and the young. Both species are good at interpreting the moods of others in the group, whether through facial expression or other forms of body language.

Legend acknowledges these shared characteristics in stories of children suckled by wolves. The other side of the same coin, in real life, means that wolf cubs adapt easily to life among humans.

For mutual benefit

Humans and wolves are competing for the same prey, but there are advantages for both in teaming up. For the wolf, human ingenuity and the use of weapons mean a share in a greater number of kills - and perhaps even an occasional taste of larger victims, such as mammoth. For humans, the wolf's speed and ferocity is equivalent to a new weapon.

The partnership is natural. So, undoubtedly, is how it first comes about. People love to nurture any abandoned young animal, and a wolf cub is well adapted to learn the rules of a hierarchical human society (in which its place will be low). From this partnership all dogs derive. Unbelievable though it seems, every single breed of dog is descended from wolves.

For a species to become domesticated, it must be willing to breed in man's company. 'Breed in captivity', the more usual phrase, implies a simple case of exploitation. The reality is more complex. In terms of survival, those species which have developed a relationship with man have far outstripped their wild cousins.

The most numerous large mammals, apart from humans, are cows, sheep, goats, pigs, horses and dogs. Domestic cats easily outnumber their wild equivalents, as do chickens and turkeys. The domestication of animals is based on an ancient contract, with benefits on both sides, between man and the ancestors of the breeds familiar to us today.

Dogs: from 12,000 years ago

The earliest known evidence of a domesticated dog is a jawbone found in a cave in Iraq and dated to about 12,000 years ago. It differs from a wolf in that it has been bred to have a smaller jaw and teeth. Selective breeding affects a species quite rapidly, and is a natural process for man to initiate - probably at first by accident rather than intention. A particular puppy in a litter is favoured because it has an attractive coat, barks well, is unusually friendly or obedient, noticeably large or small.

This is the dog which is kept and in its turn has puppies. Its desirable characteristics become perpetuated.

Images in Egyptian paintings, Assyrian sculptures and Roman mosaics reveal that by the time of these civilizations there are many different shapes and sizes of dog. To use the word 'breed' may be anachronistic, though there is evidence that a dog very like the present-day Pekingese (almost as far as one can get from a wolf) exists in China by the 1st century AD.

By that time Roman ladies also have lap dogs; their warmth is believed to be a cure for stomach ache. A Roman writer of the period gives similarly practical reasons for selecting the colour of a dog: shepherds' dogs should be white (to distinguish them from wolves in the dark) but a farmyard dog should have a black coat (to frighten thieves).

Sheep and goats, cattle and pigs: 9000-7000 BC

The first animals known to have been domesticated as a source of food are sheep in the Middle East. The proof is the high proportion of bones of one-year-old sheep discarded in a settlement at Shanidar, in what is now northern Iraq. Goats follow soon after, and these two become the standard animals of the nomadic pastoralists

- tribes which move all year long with their flocks, guided by the availability of fresh grass.

Cattle and pigs, associated more with settled communities, are domesticated slightly later - but probably not long after 7000 BC. The ox may first have been bred by humans in western Asia. The pig is probably first domesticated in China.

The first reason for herding sheep and goats, or keeping cattle and pigs in the village, is to secure a regular supply of fresh meat. The hunter is dependent on the luck of the chase; if more animals are killed than can be immediately consumed, meals from the surplus will be increasingly unpleasant as the days go by. The herdsman, by contrast, has a living larder always to hand and a supply of dairy products as well.

These animals also provide for almost every other need of neolithic man. While they are alive, they produce dung to manure the crops. When they are dead, leather and wool for garments; horn and bone for sharp points, of needles or arrows; fat for tallow candles; hooves for glue.

Draught animals: from 4000 BC

Of the four basic farm animals, cattle represent the most significant development in village life. Not only does the cow provide much more milk than its own offspring require, but the brute strength of the ox is an unprecedented addition to man's muscle power.

From about 4000 BC oxen are harnessed and put to work. They drag sledges and, somewhat later, ploughs and wheeled wagons (an almost simultaneous innovation in the Middle East and in Europe). The plough immeasurably increases the crop of wheat or rice. The wagon enables it to be brought home from more distant fields.

India and southeast Asia use another version of the domesticated ox, well adapted to hot wet conditions - the water buffalo. Whether dragging a plough-like tool through a flooded field or hauling a cart on a dry track, the buffalo is ideally suited to the role of a farm animal in rice-growing areas. Like other members of the ox family, it also provides a good supply of milk.

The buffalo is first domesticated somewhere in the near-tropical regions of Asia. Precisely where or when is not known, but buffaloes feature as domestic animals on the seals of the Indus civilization.

Cats: from before 3000 BC

Apart from dogs, cats are the only domesticated animals to dwell indoors with humans. It is also the only one which is solitary in the wild, as opposed to living in packs, herds or flocks. As a result the cat has been able to take what it wants from man (food, shelter, play) and to pay its dues in return (pest control) without losing contact with its original identity.

Cats have remained closer than other domesticated animals to their wild cousins, partly because it is so difficult to control their breeding. And they are more able than any other to fend for themselves, in the country or even in a city, if human support is withdrawn.

It is not known when cats are first domesticated. But by the time of the earliest civilization they have already acquired in the human mind a characteristic which they have never lost - the quality of mystery.

In the temples of Egypt cats are sacred animals, and are mummified in their millions. In folk stories of all nations a cat is the natural companion for people who possess an alarming second sight, such as witches.

Horses: 3000 BC

Humans acquire their most important single ally from the animal kingdom when they domesticate the horse, in about 3000 BC.

Wild horses of various kinds have spread throughout most of the world by the time human history begins. Their bones feature among the remains of early human meals, and they appear in cave paintings with other animals of the chase. Some of their earliest fossil remains have been found in America, but after arriving across the Bering Land Bridge they become extinct in that continent. They are reintroduced by European colonists in the 16th century.

A natural habitat of the wild horse is the steppes of central Asia. Here, with its ability to move fast and far, it can gallop out of harm's way and make the most of scarce grazing. And here, some 5000 years ago, humans first capture, tame and breed the horse. The original purpose, as with cattle, is to acquire a reliable source of meat and subsequently milk. But then, in a crucial development, tribesmen discover that they have at their disposal a means of transport.

With a horse beneath him, man's ability to move is improved out of all recognition. The next comparable moment in the story of human speed does not arrive for another 5000 years - with steam trains.

The first domesticated horses are of a size which we would describe as ponies. Horses of this kind were still living in the wild in Mongolia until quite recent times. Discovered there in the 1870s, and named Przewalski's horse, they survive now only in zoos.

The entire range of horses known to us, from the mighty carthorse down to the smallest ponies, is the result of human breeding. Other wild breeds, now extinct, have been added to the stock. One such example is the tarpan, which was the native breed in Europe.

Asses: 3000 BC

At much the same time as the wild horse is being domesticated in the region of the Black Sea and the Caspian, its cousin the ass or donkey (a member of the same equus family) is tamed in Egypt. At this time the donkey appears to have roamed wild in northeast Africa and up through the Fertile Crescent into Mesopotamia.

So both horse and the ass, from north and from south, become available to two of the earliest civilizations - in Mesopotamia and Egypt.

Silk moths: 3000 BC

In China an indigenous silk moth is co-opted for man's purposes. Bombyx mori is still the only insect to have been fully domesticated (in the sense that, unlike the bee, it cannot live in the wild and is not known in a wild form). The silk moth has lost the power to fly; its caterpillar can find no mulberry leaves for itself. The species exists, and survives, only because humans like silk.

The earliest known silk from bombyx mori was found in a bamboo basket unearthed by archaeologists in China. Other pieces in the same basket were from wild silkworms. The fragments date from between 2850 and 2650 BC.

Camels: 3000-1500 BC

As beasts of burden and transport, camels occupy an important place alongside horses and donkeys. Two small members of the camel family, the llama and the alpaca of south America, are domesticated first - probably before 3000 BC. At that time both species appear to have been on the verge of extinction. Domestication by the American Indians saves them. Neither the llama nor the alpaca exists now in the wild.

The larger of the two, the llama, is primarily a beast of burden, while the shaggy alpaca is valuable for its wool. Neither animal is strong enough to pull a plough or drag a cart - two important steps in the story of civilization which are denied to the early Americans.

In the parched regions of north Africa and Asia two different species of camel become the most important beasts of burden - the single-humped Arabian camel (in north Africa, the Middle East, India) and the double-humped Bactrian camel (central Asia, Mongolia). Both are well adapted to desert conditions. They can derive water, when none is available elsewhere, from the fat stored in their humps.

It is probable that they are first domesticated in Arabia some time after 1500 BC. By about 1000 BC caravans of camels are bringing precious goods up the west coast of Arabia, linking India with the Mediterranean and Mesopotamia.

Poultry and pigeons: 2000 BC

The red jungle fowl, a member of the pheasant family, lives in the forests and bamboo jungles of India and southeast Asia. The male makes an impressive crowing sound and is dignified by a comb on his head and wattles under his beak. Jungle fowl of this kind are captured and kept for their eggs and their flesh by about 2000 BC in Asia. It is thought that all domestic poultry in the world today are descended from this one species.

At much the same period, in Egypt, pigeons are first persuaded to live and breed in the proximity of humans - again as a reliable source of protein. But some 3000 years later it is discovered that they have an extra and unusual talent. Some of them can be trained to fly home.

Elephants: 2000 BC

India is the region where elephants are first tamed, during the Indus civilization. The two species of elephant are at this time widespread - the Indian elephant throughout temperate Asia as far west as Syria, and the African elephant in regions north and south of the Sahara. (The mammoth has become extinct by the end of the last glacial period, about 10,000 years ago - partly through climatic changes and partly at the hands of human hunters.)

It is not known when elephants are first trained to take part in war, but by the 3rd century BC they are a valuable military force in both India and north Africa. An ability to learn tricks also makes the elephant a performing animal, popular in the arena of the Roman circus.

The honey of the bee: before the 6th century BC

No doubt hunter-gatherers, when they find the honey of bees in a hollow tree, often risk a sting for the pleasure of sweetness. The story of beekeeping can be described as the search for safer and more convenient ways of robbing a bees' nest.

The turning point in the domestication of the bee is the discovery that a swarm of bees can be coaxed into a specific nest - one designed by man for his own convenience in collecting the honey, and with it the useful substance of beeswax.

It is not known when the beehive is first developed, but the Greeks in classical times use a design which for centuries remains standard in much of Europe. Known as a skep, it is a dome constructed from a continuous coil of woven straw - looking much like an upturned basket. It stands on a wooden platform with a hole in, through which the bees enter.

The disadvantage of such a system is that the removal of the honey involves disturbing the nest of the bees. From the 17th century, when wooden hives come into use, extra chambers are added for the collection of honey. But the major improvement in beekeeping techniques is the achievement of a 19th-century clergyman, L.L. Langstroth.

Rabbits: from the 1st century BC

Since Roman times, if not before, people have encouraged rabbits to breed in captivity for the sake of their meat, and have then regretted doing so because of the animal's ability to burrow to freedom and eat the crops. The only safe place to keep rabbits is on an island. (Almost every island of the world has rabbits on, brought by humans to establish a living larder for passing ships.)

Rabbits are inaccessible in their burrows, so man domesticates a species of polecat (in the form of the ferret) to flush them out. As early as the 1st century AD Pliny describes the use of ferrets in the Balearic islands, as the inhabitants struggle to control the rabbits (see Pliny and the ferrets).

The turkey: from the 14th century

He turkey is indigenous to central and north America. It is kept as a domestic fowl by the Aztecs in Mexico from the 14th century, and no doubt has been domesticated considerably earlier by their predecessors.

Turkeys are brought to Europe by the Spanish in the early 16th century. They become popular throughout the continent, and in the 17th century are taken back across the Atlantic by English settlers. Interbred with wild turkeys of north America, they develop into the breeds known there today.

The bee space: 1851

An American clergyman, the Rev. L.L. Langstroth, puts to practical use the very precise habits of bees. Other beekeepers in the 19th century have begun putting into the hives removable frames, in which the bees will build their honeycombs, but the bees clog the spaces up with wax and honey. In 1851 Langstroth discovers that if the frames are an exact distance apart (a little more than a quarter of an inch), the workers will keep each honey-filled frame neat and separate.

A gap which is larger or smaller does not have the same effect. The 'bee space' becomes an established principle, and the Langstroth hive is still the type used by beekeepers today.

The ostrich: 19th - 20th century

In addition to the standard domesticated animals, many others have been kept or are now kept by humans for a wide range of purposes.

A good example is the ostrich. In the late 19th-century the fashion for ostrich feathers, in hats or fans, causes farmers to domesticate and breed this largest of birds. In the late 20th-century there is more interest in it for its meat.

The hamster: 1930

In the ongoing story of domestication, the remarkable case of the golden hamster is a good detail for a closing chapter. In 1930 a female hamster with twelve young is captured at Aleppo in Syria. Taken to the laboratory at the Hebrew university in Jerusalem, they are bred for use in experiments. Each female has several litters a year, so the numbers rapidly grow. Other laboratories secure their own supply from Jerusalem. Then somebody notices that this little animal - chubby, with puffed out cheeks and soft fur - has considerable charm as a pet for children.

Half a century on the golden hamster is a common domesticated animal. Every single hamster in capitivity descends from that first Syrian litter.

Degrees

The boundaries between surviving wild populations and domestic clades can be vague. A classification system that can help solve this confusion surrounding animal populations might be set up on a spectrum of increasing domestication:

- Wild: These populations experience their full life cycles without deliberate human intervention.
- Raised in Captivity/Captured from Wild (in zoos, botanical gardens, or for human gain): These populations are nurtured by humans but (except in zoos) not normally bred under human control. They remain as a group essentially indistinguishable in appearance or behaviour from their wild counterparts. Examples include Asian elephants, animals such as sloth bears and cobras used by showmen in India, and animals such as Asian black bears (farmed for their bile), and zoo animals, kept in captivity as examples of their species. (It should be noted that zoos and botanical gardens sometimes exhibit domesticated or feral animals and plants such as camels, mustangs, and some orchids)
- Raised commercially (captive or semidomesticated): These populations are ranched or farmed in large numbers for food, commodities, or the pet trade, commonly breed in captivity, but as a group are not substantially altered in appearance or behavior from their wild cousins. Examples include the ostrich, various deer, alligator, cricket, pearl oyster, raptors used in falconry and ball python. (These species are sometimes referred to as partially domesticated.)
- Domesticated: These populations are bred and raised under human control for many generations and are substantially altered as a group in appearance or behaviour. Examples include sweet potato, garlic, pigs, ferrets, turkeys, canaries, domestic pigeons, budgerigars, goldfish, koi carp, silkworms, dogs,

cats, sheep, cattle, chickens, llamas, guinea pigs, laboratory mice, horses, goats and (silver) foxes.

This classification system does not account for several complicating factors: genetically modified organisms, feral populations, and hybridization. Many species that are farmed or ranched are now being genetically modified. This creates a unique category because it alters the organisms as a group but in ways unlike traditional domestication. Feral organisms are members of a population that was once raised under human control, but is now living and multiplying outside of human control. Examples include mustangs. Hybrids can be wild, domesticated, or both: a liger is a hybrid of two wild animals, a mule is a hybrid of two domesticated animals, and a beefalo is a cross between a wild and a domestic animal.A great difference exists between a tame animal and a domesticated animal. The term "domesticated" refers to an entire species or variety while the term "tame" can refer to just one individual within a species or variety. Humans have tamed many thousands of animals that have never been truly domesticated. These include the elephant, giraffes, and bears. There is debate over whether some species have been domesticated or just tamed. Some state that the elephant has been domesticated, while others argue the cat has never been. Dividing lines include whether a specimen born to wild parents would differ in appearance or behavior from one born to domesticated parents. For instance a dog is certainly domesticated because even a wolf (genetically share a common ancestor with all dogs) raised from a pup would be very different from a dog, in both appearance and behaviour. Similar problems of definition arise when domesticated cats go feral.

Limits

Selection of animals for visible "desirable" traits may make them unfit in other, unseen, ways. The consequences for the captive and domesticated animals were reduction in size, piebald color, shorter faces with smaller and fewer teeth, diminished horns, weak muscle ridges, and less genetic variability. Poor joint definition, late fusion of the limb bone epiphyses with the diaphyses, hair changes, greater fat accumulation, smaller brains, simplified behavior patterns, extended immaturity, and more pathology are a few of the defects of domestic animals. All of these changes have been documented in direct observations of the rat in the 19th century, by archaeological evidence, and confirmed by animal breeders in the 20th century.

One side effect of domestication has been zoonotic diseases. For example, cattle have given humanity various viral poxes, measles, and tuberculosis; pigs and ducks have given influenza; and horses have given the rhinoviruses. Humans share over sixty diseases with dogs. Many parasites also have their origins in domestic animals. The advent of domestication resulted in denser human populations which provided ripe conditions for pathogens to reproduce, mutate, spread, and eventually find a new host in humans.

DATES AND PLACES

Since the process of domestication inherently takes many generations over a long period of time, and the spread of breed and husbandry techniques is also slow, it is not meaningful to give a single "date of domestication". However, it is believed that the first attempt at domestication of both animals and plants were made in the Old World by peoples of the Mesolithic Period. The tribes that took part in hunting and gathering wild edible plants, started to make attempts to domesticate dogs, goats, and possibly sheep, which was as early as 9000 BC. However, it was not until the Neolithic Period that primitive agriculture appeared as a form of social activity, and domestication was well under way. The great majority of domesticated animals and plants that still serve humans were selected and developed during the Neolithic Period, a few other examples appeared later. The rabbit for example, was not domesticated until the Middle Ages, while the sugar beet came under cultivation as a sugar-yielding agricultural plant in the 19th century. As recently as the 20th century, mint became an object of agricultural production, and animal breeding programs to produce high-quality fur were started in the same time period.

The methods available to estimate domestication dates introduce further uncertainty, especially when domestication has occurred in the distant past. So the dates given here should be treated with caution; in some cases evidence is scanty and future discoveries may alter the dating significantly.

Dates and places of domestication are mainly estimated by archaeological methods, more precisely archaeozoology. These methods consist of excavating or studying the results of excavation in human prehistorical occupation sites. Animal remains are dated with archaeological methods, the species they belong to is determined, the age at death is also estimated, and if possible the form they had, that is to say a possible domestic form. Various other clues are taken advantage of, such as slaughter or cutting marks. The aim is to determine if they are game or raised animal, and more globally the nature of their relationship with humans. For example the skeleton of a cat found buried close to humans is a clue that it may have been a pet cat. The age structure of animal remains can also be a clue of husbandry, in which animals were killed at the optimal age.

New technologies and especially mitochondrial DNA, which are simple DNA found in the mitochondria that determine its function in the cell provide an alternative angle of investigation, and make it possible to reestimate the dates of domestication based on research into the genealogical tree of modern domestic animals.

It is admitted for several species that domestication occurred in several places distinctly. For example, research on mitochondrial DNA of the modern cattle Bos taurus supports the archaeological assertions of separate domestication events in Asia and Africa. This research also shows that Bos taurus and Bos indicus haplotypes are all descendants of the extinct wild ox Bos primigenius. However, this does not rule out later crossing inside a species; therefore it appears useless to look for a separate wild ancestor for each domestic breed.

The first animal to be domesticated appears to have been the dog, in the Upper Paleolithic era. This preceded the domestication of other species by several millennia. In the Neolithic a number of important species such as the goat, sheep, pig and cow were domesticated, as part of the spread of farming with characterises this period. The goat, sheep and pig in particular were domesticated independently in the Levant and Asia.

There is early evidence of beekeeping, in the form of rock paintings, dating to 13,000 BC.

Recent archaeological evidence from Cyprus indicates domestication of a type of cat by perhaps 9500 BC.

The earliest secure evidence of horse domestication, bit wear on horse molars at Dereivka in Ukraine, dates to around 4000 BC. The unequivocal date of domestication and use as a means of transport is at the Sintashta chariot burials in the southern Urals, c. 2000 BC. Local equivalents and smaller species were domesticated from the 26th century BC.

The availability of both domesticated vegetable and animal species increased suddenly following the voyages of Christopher Columbus and the contact between the Eastern and Western Hemispheres. This is part of what is referred to as the Columbian Exchange.

Animal husbandry

Animal husbandry is the dynamic set of relationships between humans and animals, in which qualities considered to be advantageous to humans are emphasized and further developed. The term can refer to the practice of selectively breeding and raising livestock to promote desirable traits in animals for utility, sport, pleasure, or research, but also refers to the efficient exploitation of a species in agriculture in a way that benefits all species involved in the relationship.

HISTORY OF BREEDING

Animal husbandry has been practiced for thousands of years, since the first domestication of animals. Selective breeding for desired traits was first established as a scientific practice by Robert Bakewell during the British Agricultural Revolution in the 18th century. One of his most important breeding programs was with sheep. Using native stock, he was able to quickly select for large, yet fine-boned sheep, with long, lustrous wool. The Lincoln Longwool was improved by Bakewell and in turn the Lincoln was used to develop the subsequent breed, named the New (or Dishley) Leicester. It was hornless and had a square, meaty body with straight top lines. These sheep were exported widely and have contributed to numerous modern breeds.

Under his influence, English farmers began to breed cattle for use primarily as beef for consumption - (previously, cattle were first and foremost bred for pulling ploughs as oxen). Long-horned heifers were crossed with the Westmoreland bull to eventually create the Dishley Longhorn. Over the following decades, farm animals increased dramatically in size and quality. In 1700, the average weight of a bull sold for slaughter was 370 pounds (168 kg). By 1786, that weight had more than doubled to 840 pounds (381 kg).

Animal herding professions specialized in the 19th century to include the cowboys of the United States and Canada, charros and vaqueros of Mexico, gauchos and huasos of South America, and the farmers and stockmen of Australia.

In more modern times herds are tended on horses, all-terrain vehicles, motorbikes, four-wheel drive vehicles, and helicopters, depending on the terrain and livestock concerned. Today, herd managers often oversee thousands of animals and many staff. Farms, stations and ranches may employ breeders, herd health specialists, feeders, and milkers to help care for the animals.

Breeding techniques

Techniques such as artificial insemination and embryo transfer are frequently used today, not only as methods to guarantee that females breed regularly but also to help improve herd genetics. This may be done by transplanting embryos from high-quality females into lower-quality surrogate mothers - freeing up the higher-quality mother to be reimpregnated. This practice vastly increases the number of offspring which may be produced by a small selection of the best quality parent animals. On the one hand, this improves the ability of the animals to convert feed to meat, milk, or fiber more efficiently, and improve the quality of the final product. On the other, it decreases genetic diversity, increasing the severity of disease outbreaks among other risks.

History in Europe

The semi-natural, unfertilized pastures formed by traditional agricultural methods in Europe, were managed and maintained by the grazing and mowing of livestock. Because the ecological impact of this land management strategy is similar to the impact of a natural disturbance, the agricultural system will share many beneficial characteristics with a natural habitat including the promotion of biodiversity. This strategy is declining in the European context due to the intensification of agriculture, and the mechanized chemical based methods that became popular during and following the industrial revolution.

Sustainability

In organic and biodynamic farming methods, animals play an important role in achieving closed or sustainable system by providing multiple functions to the agricultural system. In the Balinese context, for example, ducks are herded through rice paddies at particular stages of the growing season. The ducks eat insects and weeds that inhibit the growth of seedlings, and prevent the farmer from using herbicides or pesticides. Their droppings contribute nitrogen to the soil as a natural source of fertilizer, and the movement of their feet and bills in the water of the terrace ecosystem increases the oxygen content of the soil thus increasing nutrient availability.

Climate change

Due to the significant contribution of agriculture to the emissions of non-CO2 greenhouse gases methane and nitrous oxide, the relationship between humans and livestock is being analyzed for its potential to help mitigate climate change. Strategies for the mitigation include optimizing the use of gas produced from manure for energy production (biogas).

Animal husbandry in India

A large number of farmers in India depend on animal husbandry for their livelihood. In addition to supplying milk, meat, eggs,wool and hides, animals, mainly bullocks, are the major source of power for both farmers and drayers. Thus, animal husbandry plays an important role in the rural economy. The gross value of output from this sector was 358 billion (US$5.5 billion) in FY 1989, an amount that constituted about 25 percent of the total agricultural output of 1.4 trillion (US$21.4 billion).Production

In FY 1992, India had approximately 25 percent of the world's cattle, with a collective herd of 193 million head. India also had 110 million goats, 75 million water buffalo, 44 million sheep, and 10 million pigs. Milk production in FY 1990 was estimated to have reached 53.5 million tons, and egg production had reached a level of 23.3 billion eggs. Dairy farming provided supplementary employment and an additional source of income to many small and marginal farmers. The National Dairy Development Board was established in 1965 under the auspices of Operation Flood at Anand, in Gujarat, to promote, plan, and organize dairy development through cooperatives; to provide consultations; and to set up dairy plants, which were then turned over to the cooperatives. There were more than 63,000 Anand-style dairy cooperative societies with some 7.5 million members in the early 1990s. The milk produced and sold by these farmers brought 320 million (US$4.9 million) a day, or more than 10 trillion (US$153.0 billion) a year. The increase in milk production permitted India to end imports of powdered milk and milk-related products. In addition, 30,000 tons of powdered milk were exported annually to neighboring countries. There was about 50000 cows in India.Today, India has the world's largest dairy herd (composed of cows and buffaloes), at over 304 million strong, and stands first in milk production, with 112.5 million tonnes of milk produced in 2009-2010. India is also the third largest egg-producer in the world, at over 180 million eggs being produced every day or 65.7 billion eggs for the year 2011-12, and the world's sixth largest producer of poultry meat. While the majority of India's animal products are consumed domestically, exports are growing. India is the top global exporter of buffalo meat, and is also the fourth largest exporter of soybean meal, an important ingredient in commercial feed for farmed animals. In addition, India's leading poultry producers, including Suguna, Venky's, and the Amrit Group, are increasing sales to countries in other parts of Asia and the Middle East. International investment is also expanding. In 2008, U.S.-based Tyson Foods acquired a 51 percent stake in Godrej, an Indian conglomerate that is a major producer of animal feeds and poultry, and the International Finance Corporation (IFC) has an equity stake in Suguna. An estimated 200 million egg-laying hens are kept in battery cages, and more than half of the 2-billion-plus "meat" chickens produced each year are factory-farmed. Vertical integration and contract farming for poultry and eggs is also widespread. While the milk and cheese sectors still include many small-scale producers and cows and buffaloes in extensive systems the number of large, factory-style dairy operations is increasing.

Operation Flood

Operation Flood, the world's largest integrated dairy development program, attempted to establish linkages between rural milk producers and urban consumers by organizing farmer-owned and -managed dairy cooperative societies. In the early 1990s, the program was in its third phase and was receiving financial assistance from the World Bank and commodity assistance from the European Economic Community. At that time, India had more than 64,000 dairy cooperative societies, with close to 7.7 million members. These cooperatives established a daily processing capacity of 15.5 million liters of whole milk and 727 tons of milk powder.

10

Introduction to Microbiology

Microbiology is the study of microscopic organisms, either unicellular (single cell), multicellular (cell colony), or acellular (lacking cells). Microbiology encompasses numerous sub-disciplines including virology, mycology, parasitology, and bacteriology.

Eukaryotic microorganisms posses membrane-bound cell organelles and include fungi and protists, whereas prokaryotic organisms-which all are microorganisms-are conventionally classified as lacking membrane-bond organelles and include eubacteria and archaebacteria. Microbiologists traditionally relied on culture, staining, and microscopy. However, less than 1% of the microorganisms present in common environments can be cultured in isolation using current means. Microbiologists often rely on extraction or detection of nucleic acid, either DNA or RNA sequences.

Viruses have been variably classified as organisms, as they have been considered either as very simple microorganisms or very complex molecules. Prions, never considered microorganisms, have been investigated by virologists, however, as the clinical effects traced to them were originally presumed due to chronic viral infections, and virologists took search-discovering "infectious proteins".

As an application of microbiology, medical microbiology is often introduced with medical principles of immunology as microbiology and immunology. Otherwise, microbiology, virology, and immunology as basic sciences have greatly exceeded the medical variants, applied sciences.Branches

The branches of microbiology can be classified into pure and applied sciences. Microbiology can be also classified based on taxonomy, in the cases of bacteriology, mycology, protozoology, and phycology. There is considerable overlap between the specific branches of microbiology with each other and with other disciplines, and certain aspects of these branches can extend beyond the traditional scope of microbiology.

PURE MICROBIOLOGY

Taxonomic arrangement

- Bacteriology: The study of bacteria.
- Mycology: The study of fungi.
- Protozoology: The study of protozoa.
- Phycology (or algology): The study of algae.
- Parasitology: The study of parasites.
- Immunology: The study of the immune system.
- Virology: The study of viruses.

- Nematology:The study of the nematodes
- Microbiology:The study of microbes.

Integrative arrangement

- Microbial cytology: The study of microscopic and submicroscopic details of microorganisms.
- Microbial physiology: The study of how the microbial cell functions biochemically. Includes the study of microbial growth, microbial metabolism and microbial cell structure.
- Microbial ecology: The relationship between microorganisms and their environment.
- Microbial genetics: The study of how genes are organized and regulated in microbes in relation to their cellular functions. Closely related to the field of molecular biology.
- Cellular microbiology: A discipline bridging microbiology and cell biology.
- Evolutionary microbiology: The study of the evolution of microbes. This field can be subdivided into:
    - o Microbial taxonomy: The naming and classification of microorganisms.
    - o Microbial systematics: The study of the diversity and genetic relationship of microorganisms.
- Generation microbiology: The study of those microorganisms that have the same characters as their parents.
- Systems microbiology: A discipline bridging systems biology and microbiology.
- Molecular microbiology: The study of the molecular principles of the physiological processes in microorganisms.

Other

- Nano microbiology: The study of those microorganisms on nano level.
- Exo microbiology (or Astro microbiology): The study of microorganisms in outer space (see: List of microorganisms tested in outer space)
- Biological agent: The study of those microorganisms which are being used in weapon industries.

Applied microbiology

- Medical microbiology: The study of the pathogenic microbes and the role of microbes in human illness. Includes the study of microbial pathogenesis and epidemiology and is related to the study of disease pathology and immunology.
- Pharmaceutical microbiology: The study of microorganisms that are related to the production of antibiotics, enzymes, vitamins,vaccines, and other pharmaceutical products and that cause pharmaceutical contamination and spoil.
- Industrial microbiology: The exploitation of microbes for use in industrial processes. Examples include industrial fermentation and wastewater treatment. Closely linked to the biotechnology industry. This field also includes brewing, an important application of microbiology.
- Microbial biotechnology: The manipulation of microorganisms at the genetic and molecular level to generate useful products.
- Food microbiology and Dairy microbiology: The study of microorganisms causing food spoilage and foodborne illness. Using microorganisms to produce foods, for example by fermentation.

- Agricultural microbiology: The study of agriculturally relevant microorganisms. This field can be further classified into the following:
  - o Plant microbiology and Plant pathology: The study of the interactions between microorganisms and plants and plant pathogens.
  - o Soil microbiology: The study of those microorganisms that are found in soil.
- Veterinary microbiology: The study of the role of microbes in veterinary medicine or animal taxonomy.
- Environmental microbiology: The study of the function and diversity of microbes in their natural environments. This involves the characterization of key bacterial habitats such as the rhizosphere and phyllosphere, soil and groundwater ecosystems, open oceans or extreme environments (extremophiles). This field includes other branches of microbiology such as:
  - o Microbial ecology
  - o Microbially mediated nutrient cycling
  - o Geomicrobiology
  - o Microbial diversity
  - o Bioremediation
- Water microbiology (or Aquatic microbiology): The study of those microorganisms that are found in water.
- Aeromicrobiology (or Air microbiology): The study of airborne microorganisms.

Benefits

While some fear microbes due to the association of some microbes with various human illnesses, many microbes are also responsible for numerous beneficial processes such as industrial fermentation (e.g. the production of alcohol, vinegar and dairy products), antibiotic production and as vehicles for cloning in more complex organisms such as plants. Scientists have also exploited their knowledge of microbes to produce biotechnologically important enzymes such as Taq polymerase, reporter genes for use in other genetic systems and novel molecular biology techniques such as the yeast two-hybrid system.

Bacteria can be used for the industrial production of amino acids. Corynebacterium glutamicum is one of the most important bacterial species with an annual production of more than two million tons of amino acids, mainly L-glutamate and L-lysine.

A variety of biopolymers, such as polysaccharides, polyesters, and polyamides, are produced by microorganisms. Microorganisms are used for the biotechnological production of biopolymers with tailored properties suitable for high-value medical application such as tissue engineering and drug delivery. Microorganisms are used for the biosynthesis of xanthan, alginate, cellulose, cyanophycin, poly(gamma-glutamic acid), levan, hyaluronic acid, organic acids, oligosaccharides and polysaccharide, and polyhydroxyalkanoates.

Microorganisms are beneficial for microbial biodegradation or bioremediation of domestic, agricultural and industrial wastes and subsurface pollution in soils, sediments and marine environments. The ability of each microorganism to degrade toxic waste depends on the nature of each contaminant. Since sites typically have multiple pollutant types, the most effective approach to microbial biodegradation is to use a mixture of bacterial and fungal species and strains, each specific to the biodegradation of one or more types of contaminants.

Symbiotic microbial communities are known to confer various benefits to their human and animal host's health including aiding digestion, production of beneficial vitamins and amino acids, and suppression of pathogenic microbes. Some benefit may be conferred by consuming fermented foods, probiotics (bacteria potentially beneficial to the digestive system) and/or prebiotics (substances consumed to promote the growth of probiotic microorganisms). The ways the microbiome influences human and animal health, as well as methods to influence the microbiome are active areas of research.

Research has suggested that microorganisms could be useful in the treatment of cancer. Various strains of non-pathogenic clostridia can infiltrate and replicate within solid tumors. Clostridial vectors can be safely administered and their potential to deliver therapeutic proteins has been demonstrated in a variety of preclinical models.

HISTORY

Ancient

The existence of microorganisms was hypothesized for many centuries before their actual discovery. The existence of unseen microbiological life was postulated by Jainism which is based on Mahavira's teachings as early as 6th century BCE. Paul Dundas notes that Mahavira asserted existence of unseen microbiological creatures living in earth, water, air and fire. Jain scriptures also describe nigodas which are sub-microscopic creatures living in large clusters and having a very short life and are said to pervade each and every part of the universe, even in tissues of plants and flesh of animals. The Roman Marcus Terentius Varro made references to microbes when he warned against locating a homestead in the vicinity of swamps "because there are bred certain minute creatures which cannot be seen by the eyes, which float in the air and enter the body through the mouth and nose and there by cause serious diseases."

In 1546 Girolamo Fracastoro proposed that epidemic diseases were caused by transferable seedlike entities that could transmit infection by direct or indirect contact, or vehicle transmission.

However, early claims about the existence of microorganisms were speculative, and not based on microscopic observation. Actual observation and discovery of microbes had to await the invention of the microscope in the 17th century.

Modern

In 1676, Anton van Leeuwenhoek observed bacteria and other microorganisms, using a single-lens microscope of his own design. While Van Leeuwenhoek is often cited as the first to observe microbes, Robert Hooke made the first recorded microscopic observation, of the fruiting bodies of molds, in 1665. The first observation of microbes using a microscope is generally credited to the Dutch draper and haberdasher, Antonie van Leeuwenhoek, who lived for most of his life in Delft, Holland. It has, however, been suggested that a Jesuit priest called Athanasius Kircher was the first to observe micro-organisms. He was among the first to design magic lanterns for projection purposes, so he must have been well acquainted with the properties of lenses. One of his books contains a chapter in Latin, which reads in translation - 'Concerning the wonderful structure of things in nature, investigated by Microscope. Here, he wrote 'who would believe that vinegar and milk abound with an innumerable multitude of worms.' He also noted that putrid material is full of innumerable creeping animalcule.

These observations antedate Robert Hooke's Micrographia by nearly 20 years and were published some 29 years before van Leeuwenhoek saw protozoa and 37 years before he described having seen bacteria.The field of bacteriology (later a subdiscipline of microbiology) was founded in the 19th century by Ferdinand Cohn, a botanist whose studies on algae and photosynthetic bacteria led him to describe several bacteria including Bacillus and Beggiatoa. Cohn was also the first to formulate a scheme for the taxonomic classification of bacteria and discover spores. Louis Pasteur and Robert Koch were contemporaries of Cohn's and are often considered to be the father of microbiology and medical microbiology, respectively. Pasteur is most famous for his series of experiments designed to disprove the then widely held theory of spontaneous generation, thereby solidifying microbiology's identity as a biological science. Pasteur also designed methods for food preservation (pasteurization) and vaccines against several diseases such as anthrax, fowl cholera and rabies. Koch is best known for his contributions to the germ theory of disease, proving that specific diseases were caused by specific pathogenic micro-organisms. He developed a series of criteria that have become known as the Koch's postulates. Koch was one of the first scientists to focus on the isolation of bacteria in pure culture resulting in his description of several novel bacteria including Mycobacterium tuberculosis, the causative agent of tuberculosis.

While Pasteur and Koch are often considered the founders of microbiology, their work did not accurately reflect the true diversity of the microbial world because of their exclusive focus on micro-organisms having direct medical relevance. It was not until the late 19th century and the work of Martinus Beijerinck and Sergei Winogradsky, the founders of general microbiology (an older term encompassing aspects of microbial physiology, diversity and ecology), that the true breadth of microbiology was revealed. Beijerinck made two major contributions to microbiology: the discovery of viruses and the development of enrichment culture techniques. While his work on the Tobacco Mosaic Virus established the basic principles of virology, it was his development of enrichment culturing that had the most immediate impact on microbiology by allowing for the cultivation of a wide range of microbes with wildly different physiologies. Winogradsky was the first to develop the concept of chemolithotrophy and to thereby reveal the essential role played by micro-organisms in geochemical processes. He was responsible for the first isolation and description of both nitrifying and nitrogen-fixing bacteria. French-Canadian microbiologist Felix d'Herelle co-discovered bacteriophages and was one of the earliest applied microbiologists.

CLASSIFICATION OF MICROORGANISMS

Microorganisms can be found almost anywhere in the taxonomic organization of life on the planet. Bacteria and archaea are almost always microscopic, while a number of eukaryotes are also microscopic, including most protists, some fungi, as well as some animals and plants. Viruses are generally regarded as not living and therefore not considered as microbes, although the field of microbiology also encompasses the study of viruses.

Prokaryotes

Prokaryotes are organisms that lack a cell nucleus and the other membrane bound organelles. They are almost always unicellular, although some species such as myxobacteria can aggregate into complex structures as part of their life cycle.

Consisting of two domains, bacteria and archaea, the prokaryotes are the most diverse and abundant group of organisms on Earth and inhabit practically all environments where the temperature is below +140 C. They are found in water, soil, air, animals' gastrointestinal tracts, hot springs and even deep beneath the Earth's crust in rocks. Practically all surfaces that have not been specially sterilized are covered by prokaryotes. The number of prokaryotes on Earth is estimated to be around five million trillion trillion, or 5 10, accounting for at least half the biomass on Earth.

Bacteria

Almost all bacteria are invisible to the naked eye, with a few extremely rare exceptions, such as Thiomargarita namibiensis. They lack a nucleus and other membrane-bound organelles, and can function and reproduce as individual cells, but often aggregate in multicellular colonies. Their genome is usually a single loop of DNA, although they can also harbor small pieces of DNA called plasmids. These plasmids can be transferred between cells through bacterial conjugation. Bacteria are surrounded by a cell wall, which provides strength and rigidity to their cells. They reproduce by binary fission or sometimes by budding, but do not undergo meiotic sexual reproduction. However, many bacterial species can transfer DNA between individual cells by a process referred to as natural transformation. (Also see Transformation (genetics)). In nature, the development of competence for transformation is usually associated with stressful environmental conditions, and seems to be an adaptation for facilitating repair of DNA damage in recipient cells. (Also see Natural competence.) Some species form extraordinarily resilient spores, but for bacteria this is a mechanism for survival, not reproduction. Under optimal conditions bacteria can grow extremely rapidly and can double as quickly as every 20 minutes.

Archaea

Archaea are also single-celled organisms that lack nuclei. In the past, the differences between bacteria and archaea were not recognised and archaea were classified with bacteria as part of the kingdom Monera. However, in 1990 the microbiologist Carl Woese proposed the three-domain system that divided living things into bacteria, archaea and eukaryotes. Archaea differ from bacteria in both their genetics and biochemistry. For example, while bacterial cell membranes are made from phosphoglycerides with ester bonds, archaean membranes are made of ether lipids.

Archaea were originally described in extreme environments, such as hot springs, but have since been found in all types of habitats. Only now are scientists beginning to realize how common archaea are in the environment, with crenarchaeota being the most common form of life in the ocean, dominating ecosystems below 150 m in depth. These organisms are also common in soil and play a vital role in ammonia oxidation.

Eukaryotes

Most living things that are visible to the naked eye in their adult form are eukaryotes, including humans. However, a large number of eukaryotes are also microorganisms. Unlike bacteria and archaea, eukaryotes contain organelles such as the cell nucleus, the Golgi apparatus and mitochondria in their cells. The nucleus is an organelle that houses the DNA that makes up a cell's genome. DNA itself is arranged in complex chromosomes. Mitochondria are organelles vital in metabolism as they are the site of the citric acid cycle and oxidative phosphorylation. They evolved from symbiotic bacteria and retain a remnant genome. Like bacteria, plant cells have cell walls, and contain organelles such as chloroplasts in addition to the organelles in other eukaryotes. Chloroplasts produce energy from light by photosynthesis, and were also originally symbiotic bacteria.

Unicellular eukaryotes consist of a single cell throughout their life cycle. This qualification is significant since most multicellular eukaryotes consist of a single cell called a zygote only at the beginning of their life cycles. Microbial eukaryotes can be either haploid or diploid, and some organisms have multiple cell nuclei (see coenocyte).

Unicellular eukaryotes usually reproduce asexually by mitosis under favorable conditions. However, under stressful conditions such as nutrient limitations and other conditions associated with DNA damage, they tend to reproduce sexually by meiosis and syngamy. (Also see Meiosis.)

Protists

Of eukaryotic groups, the protists are most commonly unicellular and microscopic. This is a highly diverse group of organisms that are not easy to classify. Several algae species are multicellular protists, and slime molds have unique life cycles that involve switching between unicellular, colonial, and multicellular forms. The number of species of protists is unknown since we may have identified only a small proportion. Studies from 2001-2004 have shown that a large number of protist diversity exitsts in oceans, deep sea-vents, river sediment and an acidic river which suggests that a large number of eukaryotic microbial communities have yet to be discovered.

Animals

Micro-animals

Most animals are multicellular, but some are too small to be seen by the naked eye. Microscopic arthropods include dust mites and spider mites. Microscopic crustaceans include copepods and the cladocera, while many nematodes are too small to be seen with the naked eye. Another particularly common group of microscopic animals are the rotifers, which are filter feeders that are usually found in fresh water. Micro-animals reproduce both sexually and asexually and may reach new habitats as some eggs can survive harsh environments that would kill the adult animal. However, some simple animals, such as rotifers and nematodes, can dry out completely and remain dormant for long periods of time.

Fungi

The fungi have several unicellular species, such as baker's yeast (Saccharomyces cerevisiae) and fission yeast (Schizosaccharomyces pombe). Some fungi, such as the pathogenic yeast Candida albicans, can undergo phenotypic switching and grow as single cells in some environments, and filamentous hyphae in others. Fungi reproduce both asexually, by budding or binary fission, as well by producing spores, which are called conidia when produced asexually, or basidiospores when produced sexually.

Plants

The green algae are a large group of photosynthetic eukaryotes that include many microscopic organisms. Although some green algae are classified as protists, others such as charophyta are classified with embryophyte plants, which are the most familiar group of land plants. Algae can grow as single cells, or in long chains of cells. The green algae include unicellular and colonial flagellates, usually but not always with two flagella per cell, as well as various colonial, coccoid, and filamentous forms. In the Charales, which are the algae most closely related to higher plants, cells differentiate into several distinct tissues within the organism. There are about 6000 species of green algae.

11

Animal Nutrition

Animal nutrition focuses on the dietary needs of domesticated animals, primarily those in agriculture and food production.

INTRODUCTION TO ANIMAL NUTRITION

There are seven major classes of nutrients: carbohydrates, fats, fiber, minerals, protein, vitamin, and water.

These nutrient classes can be categorized as either macronutrients (needed in relatively large amounts) or micronutrients (needed in smaller quantities). The macronutrients are carbohydrates, fats, fiber, proteins, and water. The micronutrients are minerals and vitamins.

The macronutrients (excluding fiber and water) provide structural material (amino acids from which proteins are built, and lipids from which cell membranes and some signaling molecules are built) and energy. Some of the structural material can be used to generate energy internally, and in either case it is measured in joules or calories (sometimes called "kilocalories" and on other rare occasions written with a capital C to distinguish them from little 'c' calories). Carbohydrates and proteins provide 17 kJ approximately (4 kcal) of energy per gram, while fats provide 37 kJ (9 kcal) per gram., though the net energy from either depends on such factors as absorption and digestive effort, which vary substantially from instance to instance. Vitamins, minerals, fiber, and water do not provide energy, but are required for other reasons. A third class dietary material, fiber (i.e., non-digestible material such as cellulose), seems also to be required, for both mechanical and biochemical reasons, though the exact reasons remain unclear.

Molecules of carbohydrates and fats consist of carbon, hydrogen, and oxygen atoms. Carbohydrates range from simple monosaccharides (glucose, fructose, galactose) to complex polysaccharides (starch). Fats are triglycerides, made of assorted fatty acid monomers bound to glycerol backbone. Some fatty acids, but not all, are essential in the diet: they cannot be synthesized in the body. Protein molecules contain nitrogen atoms in addition to carbon, oxygen, and hydrogen. The fundamental components of protein are nitrogen-containing amino acids, some of which are essential in the sense that humans cannot make them internally. Some of the amino acids are convertible (with the expenditure of energy) to glucose and can be used for energy production just as ordinary glucose. By breaking down existing protein, some glucose can be produced internally; the remaining amino acids are discarded, primarily as urea in urine. This occurs normally only during prolonged starvation.

Other micronutrients include antioxidants and phytochemicals which are said to influence (or protect) some body systems. Their necessity is not as well established as in the case of, for instance, vitamins.

Most foods contain a mix of some or all of the nutrient classes, together with other substances such as toxins or various sorts. Some nutrients can be stored internally (e.g., the fat soluble vitamins), while others are required more or less continuously. Poor health can be caused by a lack of required nutrients or, in extreme cases, too much of a required nutrient. For example, both salt and water (both absolutely required) will cause illness or even death in too large amounts.

Carbohydrates

Carbohydrates may be classified as monosaccharides, disaccharides, or polysaccharides depending on the number of monomer (sugar) units they contain. They constitute a large part of foods such as rice, noodles, bread, and other grain-based products. Monosaccharides contain one sugar unit, disaccharides two, and polysaccharides three or more. Polysaccharides are often referred to as complex carbohydrates because they are typically long multiple branched chains of sugar units. The difference is that complex carbohydrates take longer to digest and absorb since their sugar units must be separated from the chain before absorption. The spike in blood glucose levels after ingestion of simple sugars is thought to be related to some of the heart and vascular diseases which have become more frequent in recent times. Simple sugars form a greater part of modern diets than formerly, perhaps leading to more cardiovascular disease. The degree of causation is still not clear, however.

Fat

A molecule of dietary fat typically consists of several fatty acids (containing long chains of carbon and hydrogen atoms), bonded to a glycerol. They are typically found as triglycerides (three fatty acids attached to one glycerol backbone). Fats may be classified as saturated or unsaturated depending on the detailed structure of the fatty acids involved. Saturated fats have all of the carbon atoms in their fatty acid chains bonded to hydrogen atoms, whereas unsaturated fats have some of these carbon atoms double-bonded, so their molecules have relatively fewer hydrogen atoms than a saturated fatty acid of the same length. Unsaturated fats may be further classified as monounsaturated (one double-bond) or polyunsaturated (many double-bonds). Furthermore, depending on the location of the double-bond in the fatty acid chain, unsaturated fatty acids are classified as omega-3 or omega-6 fatty acids. Trans fats are a type of unsaturated fat with trans-isomer bonds; these are rare in nature and in foods from natural sources; they are typically created in an industrial process called (partial) hydrogenation.

Many studies have shown that unsaturated fats, particularly monounsaturated fats, are best in the human diet. Saturated fats, typically from animal sources, are next, while trans fats are to be avoided. Saturated and some trans fats are typically solid at room temperature (such as butter or lard), while unsaturated fats are typically liquids (such as olive oil or flaxseed oil). Trans fats are very rare in nature, but have properties useful in the food processing industry, such as rancid resistance.

ESSENTIAL FATTY ACIDS

Most fatty acids are non-essential, meaning the body can produce them as needed, generally from other fatty acids and always by expending energy to do so. However, in humans at least two fatty acids are essential and must be included in the diet. An appropriate balance of essential fatty acids -- omega-3 and omega-6 fatty acids -- seems also important for health, though definitive experimental demonstration has been elusive. Both of these "omega" long-chain polyunsaturated fatty acids are substrates for a class of eicosanoids known as prostaglandins, which have roles throughout the human body. They are hormones, in some respects. The omega-3 eicosapentaenoic acid (EPA), which can be made in the human body from the omega-3 essential fatty acid alpha-linolenic acid (LNA), or taken in through marine food sources, serves as a building block for series 3 prostaglandins (e.g. weakly inflammatory PGE3). The omega-6 dihomo-gamma-linolenic acid (DGLA) serves as a building block for series 1 prostaglandins (e.g. anti-inflammatory PGE1), whereas arachidonic acid (AA) serves as a building block for series 2 prostaglandins (e.g. pro-inflammatory PGE 2). Both DGLA and AA can be made from the omega-6 linoleic acid (LA) in the human body, or can be taken in directly through food. An appropriately balanced intake of omega-3 and omega-6 partly determines the relative production of different prostaglandins: one reason a balance between omega-3 and omega-6 is believed important for cardiovascular health. In industrialized societies, people typically consume large amounts of processed vegetable oils, which have reduced amounts of the essential fatty acids along with too much of omega-6 fatty acids relative to omega-3 fatty acids.

The conversion rate of omega-6 DGLA to AA largely determines the production of the prostaglandins PGE1 and PGE2. Omega-3 EPA prevents AA from being released from membranes, thereby skewing prostaglandin balance away from pro-inflammatory PGE2 (made from AA) toward anti-inflammatory PGE1 (made from DGLA). Moreover, the conversion (desaturation) of DGLA to AA is controlled by the enzyme delta-5-desaturase, which in turn is controlled by hormones such as insulin (up-regulation) and glucagon (down-regulation). The amount and type of carbohydrates consumed, along with some types of amino acid, can influence processes involving insulin, glucagon, and other hormones; therefore the ratio of omega-3 versus omega-6 has wide effects on general health, and specific effects on immune function and inflammation, and mitosis (i.e. cell division).

Good sources of essential fatty acids include most vegetables, nuts, seeds, and marine oils, Some of the best sources are fish, flax seed oils, soy beans, pumpkin seeds, sunflower seeds, and walnuts.

Bibliography

- Antonio, Saltini. Storia delle scienze agrarie, 4 vols, Bologna 1984-89, ISBN 88-206-2412-5, ISBN 88-206-2413-3, ISBN 88-206-2414-1, ISBN 88-206-2415-X
- Juliet, Clutton Brock. The walking larder. Patterns of domestication, pastoralism and predation, Unwin Hyman, London 1988
- Juliet, Clutton Brock. Horse power: a history of the horse and donkey in human societies, National history Museum publications, London 1992
- Fleming G., Guzzoni M. Storia cronologica delle epizoozie dal 1409 av. Cristo sino al 1800, in Gazzetta medico-veterinaria, I-II, Milano 1871-72
- Hall S and Juliet, Clutton Brock. Two hundred years of British farm livestock, Natural History Museum Publications, London 1988
- Jules, Janick; Noller, Carl H. and Rhyker, Charles L. The Cycles of Plant and Animal Nutrition, in Food and Agriculture, Scientific American Books, San Francisco 1976
- Manger, Louis N. A History of the Life Sciences, M. Dekker, New York, Basel 2002
- Berg J, Tymoczko JL, Stryer L (2002). Biochemistry (5th ed.). San Francisco: W.H. Freeman. p. 603. ISBN 0-7167-4684-0.
- Barker, Helen M. (2002). Nutrition and dietetics for health care. Edinburgh: Churchill Livingstone. p. 17. ISBN 0-443-07021-0. OCLC 48917971
- Nelson, D. L.; Cox, M. M. (2000). Lehninger Principles of Biochemistry (3rd ed.). New York: Worth Publishing. ISBN 1-57259-153-6.
- D. E. C. Corbridge (1995). Phosphorus: An Outline of its Chemistry, Biochemistry, and Technology (5th ed.). Amsterdam: Elsevier. ISBN 0-444-89307-5.
- Lippard, S. J. and Berg, J. M. (1994). Principles of Bioinorganic Chemistry. Mill Valley, CA: University Science Books. ISBN 0-935702-73-3.
- " Hammond K. Gianola, D (1990), Advances in Statistical Methods for Genetic Improvement of Livestock (Advanced Series in Agricultural Sciences), Springer-Verlag Berlin and Heidelberg GmbH & Co. K, ISBN 3-540-50809-0
- " Massey, JW and Vogt, DW (1993), Heritability and Its Use in Animal Breeding, Department of Animal Sciences, University of Missouri
- " Mrode, R. A. (1996), Linear models for the prediction of animal breeding values, Oxon: CAB International, ISBN 0-85198-996-9
- "Cameron, N. D. (1997), Selection indices and prediction of genetic merit in animal breeding, Oxon: CAB International, ISBN 0-85199-169-6
- "Dalton, C, Willis, MB (1998), Dalton's Introduction to Practical Animal Breeding, Oxford: Blackwell Science, ISBN 0-632-04947-2

- " Bourdon, RM (2000), Understanding animal breeding, Englewood Cliffs, N.J: Prentice Hall, ISBN 0-13-096449-2
- "Newman, S, Rothschild, MF (2002), Intellectual Property Rights in Animal Breeding and Genetics, Wallingford, Oxon, UK: CABI Pub, ISBN 0-85199-641-8

Index